Nowcasting

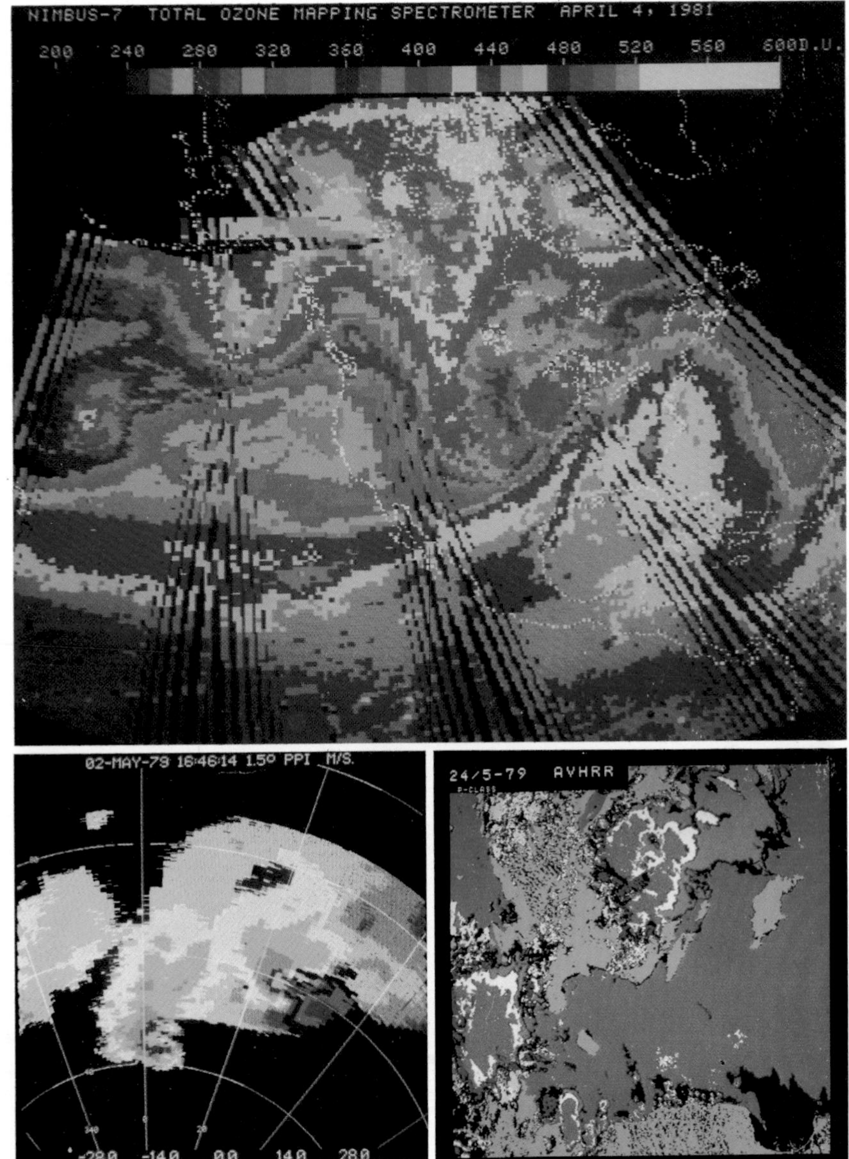

1

2

3

For caption see over

Frontispiece

(1) High-resolution plan display of total ozone, as observed by the Total Ozone Mapping Spectrometer on the Nimbus 7 satellite at local noon on 4 April 1981 over North America. Ozone amounts shown by the colour scale are in Dobson units (1000 DU = 1 atm cm). Low and high total ozone features correspond to upper tropospheric ridges and troughs, respectively. Jet streams are found where the ozone gradient is a maximum, i.e. in an arc extending from Baja California to Lake Michigan. (See Chapter 2.5 by Shapiro, Krueger and Kennedy.)

(2) Doppler velocity plan display of two tornado-producing thunderstorms in Oklahoma on 2 May 1979, obtained using a ground-based pulse Doppler radar. Range rings are at 20-km intervals. Colours represent radial velocity as given by the colour bar in m s^{-1}. One tornado, at 028°/68 km, was within a mesocyclone which had a maximum approaching velocity of 32 m s^{-1} on its west side and a maximum receding velocity of 42 m s^{-1} a few kilometres away on its east side. The second tornado was associated with a rather less intense mesocyclone centred at 005°/45 km. (See Chapter 2.2 by Wilson and Wilk.)

(3) High-resolution plan display of cloud systems, derived from an automated multispectral analysis of AVHRR satellite imagery. The picture covers southern Sweden and the Baltic. Red, cumulonimbus; light rose, nimbostratus; violet, cirrus or cirrostratus; brown, cumulus congestus; yellow, cumulus humilis; green, land surfaces; and blue, sea or lake surfaces. (This plate shows part of the larger area reproduced in black and white in Chapter 3.2 by Liljas.)

Nowcasting

edited by

K.A. BROWNING
Meteorological Office Radar Research Laboratory, Malvern, UK

ACADEMIC PRESS

A Subsidiary of Harcourt Brace Jovanovich, Publishers

LONDON NEW YORK
PARIS SAN DIEGO SAN FRANCISCO SÃO PAULO
SYDNEY TOKYO TORONTO

Academic Press Inc. (London) Ltd
24–28 Oval Road
London NW1

US edition published by
Academic Press Inc.
111 Fifth Avenue,
New York, New York 10003

British Library Cataloguing in Publication Data

Nowcasting.
1. Weather forecasting–Congresses
I. Browning, K.A.
551.6'362 QC995

ISBN 0-12-137760-1

LCCCN 82-45030

Printed in Great Britain at the Alden Press
Oxford London and Northampton

Contributors

AUSTIN, G. L.
McGill Radar Weather Observatory, Box 241, Macdonald Campus, Ste Anne de Bellevue, Canada PQ H9X 1CO

BELLON, A.
McGill Radar Weather Observatory, Box 241, Macdonald Campus, Ste Anne de Bellevue, Canada PQ H9X 1CO

BERAN, D. W.
PROFS Program Office, National Oceanic and Atmospheric Administration, Environmental Research Laboratories, Boulder, Colorado 80303, USA

BODIN, S.
Swedish Meteorological and Hydrological Institute, Box 923 S/601–19, Norrköping, Sweden

BROWN, B. G.
Department of Atmospheric Sciences, Oregon State University, Corvallis, Oregon 97331, USA

BROWNING, K. A.
Meteorological Office Radar Research Laboratory, Royal Signals and Radar Establishment, Malvern, Worcestershire WR14 3PS, UK

CARPENTER, K. M.
Meteorological Office Radar Research Laboratory, Royal Signals and Radar Establishment, Malvern, Worcestershire WR14 3PS, UK

COLLIER, C. G.
Meteorological Office Radar Research Laboratory, Royal Signals and Radar Establishment, Malvern, Worcestershire WR14 3PS, UK

FORGAN, B. W.
Australian Numerical Meteorology Research Centre, Melbourne, Victoria 3001, Australia

HITSUMA, M.
Japan Meteorological Agency, Ote-machi, Chiyoda-ku, Tokyo, Japan

KELLY, G. A. M.
Australian Numerical Meteorology Research Centre, Melbourne, Victoria 3001, Australia

KENNEDY, P. J.
National Center for Atmospheric Research,† Boulder, Colorado 80307, USA

†The National Center for Atmospheric Research is sponsored by the National Science Foundation

KRUEGER, A. J.
NASA, Goddard Space Flight Center, Greenbelt, Maryland 20771, USA

LILJAS, E.
Swedish Meteorological and Hydrological Institute, Box 923 S/601–19, Norrköping, Sweden

LITTLE, C. G.
Wave Propagation Laboratory, National Oceanic and Atmospheric Administration, Environmental Research Laboratories, Boulder, Colorado 80303, USA

MACDONALD, A. E.
PROFS Program Office, National Oceanic and Atmospheric Administration, Environmental Research Laboratories, Boulder, Colorado 80303, USA

MAKINO, Y.
Japan Meteorological Agency, Ote-machi, Chiyoda-ku, Tokyo, Japan

LE MARSHALL, J. F.
National Meteorological Analysis Centre, Bureau of Meteorology, Melbourne, Victoria 3001, Australia

MENZEL, W. P.
University of Wisconsin, Madison, Wisconsin 53706, USA

MURPHY, A. H.
Department of Atmospheric Sciences, Oregon State University, Corvallis, Oregon 97331, USA

PIELKE, R. A.*
Department of Environmental Sciences, University of Virginia, Charlottesville, Virginia 22903, USA

POWERS, P. E.
Australian Numerical Meteorology Research Centre, Melbourne, Victoria 3001, Australia

PURDOM, J. F. W.
Regional and Mesoscale Meteorology Branch, Applications Laboratory, National Oceanic and Atmospheric Administration/National Environmental Satellite Service, Colorado State University, Fort Collins, Colorado 80523, USA

SHAPIRO, M. A.
National Center for Atmospheric Research,† Boulder, Colorado 80307, USA

*Present address: Department of Atmospheric Science, Colorado State University, Fort Collins, Colorado 80523, USA

†The National Center for Atmospheric Research is sponsored by the National Science Foundation

SMITH, W. L.
Development Laboratory, National Oceanic and Atmospheric Administration/National Environmental Satellite Service, 1225 West Dayton Street, Madison, Wisconsin 53706, USA

SUOMI, V. E.
University of Wisconsin, Madison, Wisconsin 53706, USA

TARBELL, T. C.
United States Air Force, Air Force Global Weather Central, Offutt Air Force Base, Nebraska 68113, USA

TATEHIRA, R.
Japan Meteorological Agency, Ote-machi, Chiyoda-ku, Tokyo, Japan

WARNER, T. T.
Department of Meteorology, The Pennsylvania State University, University Park, Pennsylvania 16802, USA

WILK, K. E.
National Severe Storms Laboratory, Norman, Oklahoma 73069, USA

WILSON, J. W.
National Center for Atmospheric Research,† Boulder, Colorado 80307, USA

WOLCOTT, S. W.‡
Department of Meteorology, The Pennsylvania State University, University Park, Pennsylvania 16802, USA

ZHOU, F. X.
Institute for Atmospheric Physics, Beijing, China

ZIPSER, E. J.
National Center for Atmospheric Research,† Boulder, Colorado 80307, USA

†The National Center for Atmospheric Research is sponsored by the National Science Foundation

‡Present address: United States Geological Survey, Albany, New York 13301, USA

Preface

This book contains an edited selection of seventeen chapters arising out of the Nowcasting Symposium held in Hamburg on 25–28 August 1981 as part of the Third Scientific Assembly of the International Association of Meteorology and Atmospheric Physics. The original conference preprint volume, containing four times as many contributions, was published by the European Space Agency (ESA SP-165).

The term nowcasting symbolizes an observations-intensive approach to local weather forecasting, with timely use of current data, in which remote sensing plays a dominant role. Nowcasting is a rather inelegant term but it has the merit of having helped focus attention on an important growth area in meteorology. The term gained prominence in the mid-1970s with the Chesapeake Bay Region Nowcasting Experiment, which was designed to give timely and detailed weather information to the local community on the basis of frequent cloud imagery from a geostationary satellite together with radar and other data. The very word, nowcasting, evokes a vivid picture of an approach to prediction that is more than usually dependent on the description of the present state. Strictly, nowcasting is defined as a detailed description of the current weather along with forecasts obtained by extrapolation up to 2 hours ahead. However, the scope of this book is extended to cover very-short-range forecasting up to 12 hours ahead, including both linear extrapolation methods and dynamical (and other) methods which allow for development and decay.

The weather phenomena which are the subject of the nowcast are associated with mesoscale systems. The mesoscale lies between the synoptic scale and the cumulus scale: hence its name. It can be anything between a few kilometres and several hundred kilometres, with time scales between an hour and a day. Fronts, thunderstorm systems, and various kinds of local terrain-induced effects, all occur on the mesoscale.

Traditional weather forecasting is based upon widely spaced observational data (basically the synoptic radiosonde network of upper air stations). The resulting forecasts are general in nature. General forecasts for a day or more ahead have undoubtedly improved in line with the continuing development of numerical–dynamical weather prediction models. But, even though some operational models are now referred to as fine-mesh models, they do not have the resolution, initial data and/or physical realism to describe mesoscale weather systems. Thus the traditional methods have not led to parallel improvements in the quality of forecasts of local weather for the period up to 12 hours ahead. The last two decades, however, have seen considerable advances in our understanding of mesoscale weather systems, in our ability to make detailed observations on the mesoscale, and in our ability to process and communicate large volumes of data. As a result the meteorological community is in a position to implement regionally based nowcasting systems capable of providing more accurate and site-specific very-short-range forecasts. The opportunities are real but so too are the challenges in taking advantage of them.

The field of nowcasting is in a state of flux and this is not the time for a definitive treatise on the subject. Instead it is the purpose of this book to bring into sharp focus the nature and interrelationships (see block diagram) of the diverse subject areas that make up the field of nowcasting, and thereby to give its development added impetus. An attempt has been made to ensure that the individual subject areas are presented in a way that is intelligible and of value to both research and operational meteorologists, whether or not they have special expertise in the individual areas. Although the book is intended to impart a sense of the direction and the potential of nowcasting, the reader must appreciate that many of the ideas and techniques are exploratory, and beware of the false trails we may lay.

The book is in four parts. Part 1 deals with different aspects of the design of nowcasting systems, with an emphasis on the need for the components of the system (observations, analysis and forecasting, data processing and communications) to fit together as a whole. Part 2 deals with new forms of observational data for nowcasting; it illustrates what amounts almost to a revolution in mesoscale observational capability as a result of advances in remote sensing, both ground-based and from space. There are two kinds of remote sensing observations that are notably absent from Part 2 – satellite cloud imagery and conventional weather radar data. These are already well established and the challenge with them is not so much in the observational techniques themselves as in the use and interpretation of the data for nowcasting; these aspects are dealt with in Part 3 where the emphasis is on both the subjective use of the data to improve the forecaster's understanding of the mesoscale situation and on the objective interpretation and extrapolation of the observations. Finally, in Part 4, the prospects and limitations of mesoscale numerical models are discussed from the point of view of their application to very-short-range forecasting and the extent to which such models might make use of detailed nowcast data.

This book makes no claim to be comprehensive: indeed, by striving to present the nowcasting concept as succinctly as possible, many useful contributions at the Symposium have been omitted. The relative lack of material from the Tropics is also regrettable, because nowcasting methods based on remote sensing are likely to prove particularly helpful in areas rich in mesoscale events but poor in terms of existing forecasting facilities. Nevertheless, an attempt has been made to achieve a balanced selection of contributions from some of the active workers in this field and it is to be hoped therefore that the book will succeed in providing a foretaste of what can be expected from this emerging subject.

K A Browning
October 1981

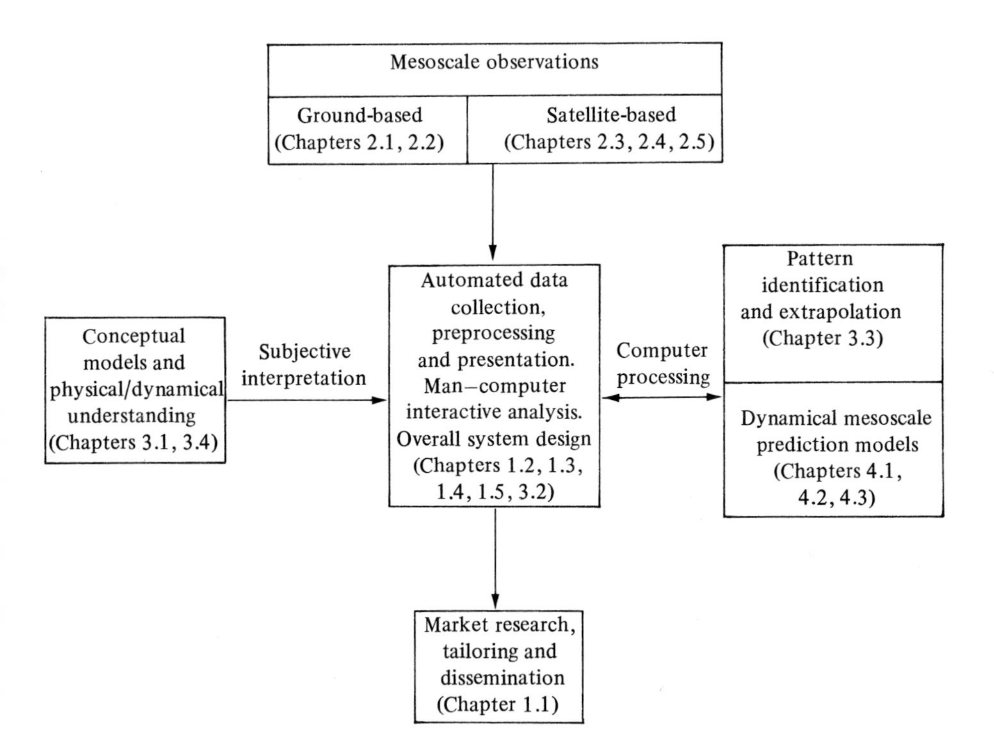

Diagram showing how the components of the nowcasting and very-short-range forecasting system are distributed amongst the chapters.

Contents

PART 1

System Design

Introduction

Any weather forecasting system, if it is to meet requirements in a cost-effective way, must be designed as a total system – all the way from the observations, through the various steps of data processing, analysis and forecasting, to the tailoring and dissemination of specific advice for real customers. In the case of nowcasting and very-short-range forecasting, spectacularly large amounts of observational data have to be processed and the resulting products distributed to the users very quickly because of the perishable nature of the information. The case for a proper systems approach is thus particularly strong, and so we have decided to begin this book with overall systems aspects. This then sets the context for consideration of the individual components in Parts 2, 3 and 4.

We start with a discussion by Murphy and Brown (Chapter 1.1) of user requirements. As they point out, there have been few soundly based economic studies of the value of weather information. Thus we have to admit at the outset that our belief in the cost-effectiveness of nowcasting systems rests more on intuition than upon quantitative evidence. Chapter 1.1 goes on to establish guidelines for the kind of studies of user requirements that would enable us to optimize the design of individual nowcasting systems, which for good meteorological and geographical as well as economic reasons will differ considerably from one region or country to another.

Beran and MacDonald (Chapter 1.2) discuss the components of the total nowcasting and very-short-range forecasting system. They then present a design methodology by means of which one may evolve an optimum system. This approach is currently being applied in the pioneering PROFS program in the United States.

Bodin (Chapter 1.3) outlines a plan for a future national forecasting system. The plan is for the new Swedish system, and it is notable for its recognition of the need for a newly structured framework to take into account the availability of new technologies and an increased emphasis on very-short-range local forecasts. Although other countries with different meteorological problems and priorities may need to evolve quite different plans, it is nevertheless instructive to consider the reasoning behind the proposed system design.

Chapter 1.4 concentrates on the observational system. Later chapters will be dealing with new forms of observational techniques; in this chapter, however, Tatehira and his colleagues describe a system consisting of fairly conventional instrumentation. The Japanese mesoscale network is one of the best equipped operational networks now in existence. It has the benefit of twenty weather radars and of the order of a thousand automatic weather stations with a spacing of about 20 km. Whilst it might be harder to justify such a network in countries less prone to severe weather, this chapter does provide a useful indication of the kind of network to which many local forecasters would aspire.

Multiple radars and statellite imagery provide some of the most important mesoscale data for nowcasting. The problem is in designing a system in which these data may be integrated with one another and with other meteorological information in such a way as to provide prompt and reliable analyses and forecasts. The rather unusual error characteristics of these data demand that special precautions be taken to ensure a credible product. Browning and Collier (Chapter 1.5) describe a total system approach being developed in the UK by means of which the forecaster can exercise his judgment effectively within the context of an otherwise highly automated system.

K. A. Browning

1.1

User Requirements for Very-short-range Weather Forecasts[1]

ALLAN H. MURPHY and BARBARA G. BROWN

1 Introduction

Studies of user requirements should play an important role in any effort to improve or enhance the usefulness of weather forecasts by introducing new observing, forecasting, or dissemination systems. Specifically, the results of such studies can provide information essential to a variety of individuals involved in designing, developing, and implementing these systems, as well as to potential users of the output of the systems. Numerous investigations related to user requirements for weather information have been undertaken in the last two decades, with the lead times of the information of interest ranging from current data to long-range forecasts. Since only nowcasts and very-short-range (i.e. 0–12h) forecasts are of concern here, however, relatively few of these studies contain directly relevant data. Moreover, the studies of interest focus on different aspects or dimensions of user requirements and employ a variety of methodologies. In addition, the quality of the information provided by these investigations varies greatly from study to study. Clearly, to obtain reliable and useful results and to provide a sound basis for decisions related to system design and development, user-requirement studies should be carefully designed and based on the most appropriate methodology available.

The primary objectives of this chapter are (a) to describe briefly currently available information on user requirements for very-short-range forecasts, (b) to identify important deficiencies in this state of knowledge, and (c) to discuss the need for future work in this area.

2 Review of some recent studies

In this section we briefly summarize the results of some recent studies of user requirements for very-short-range forecasts. For the present purposes, the

[1] Supported in part by the National Science Foundation (Division of Atmospheric Sciences) under grant ATM80-04680.

population of users is divided into five areas or sectors: (a) agriculture, (b) construction, (c) energy, (d) transportation, and (e) public safety and general public. A subsection is devoted to user needs in each sector. In describing these requirements, only the forecast-sensitive activities of users and the relevant weather variables are identified. Because user requirements for weather forecasts cannot be evaluated without proper consideration of the economic values of such forecasts (see section 3), some information on economic values of nowcasts is also provided. Subsection 2.6 consists of a brief summary and a table containing selected data on user requirements and economic values of very-short-range forecasts.

It should be emphasized that we are taking the requirements at "face value". However, as indicated above, these data suffer from several important deficiencies. The nature of these deficiencies, as well as the need for future user-requirement studies, are examined in section 3.

2.1 AGRICULTURE

The very-short-range forecast requirements of agricultural users include wind speed and direction forecasts for effective rice seeding from aircraft; wind forecasts, as well as forecasts of precipitation timing and duration, for efficient fungicide, insecticide, and herbicide spraying programs; and forecasts of the duration and intensity of precipitation to minimize damage to lettuce and other seedlings. Also required are timely forecasts of minimum temperature, dewpoint, cloud cover, and wind speed to protect citrus and deciduous fruit crops – as well as crops such as grapes, tobacco, and sugar cane – from frost; maximum temperature forecasts to reduce the adverse effects of high temperatures on grapes and other crops as well as on livestock; forecasts of humidity and cloud cover to prevent losses caused by downy mildew and other fungal diseases on deciduous fruit; and hail forecasts to minimize damage to livestock and greenhouses. Very-short-range forecast requirements during and after harvest include precipitation timing and duration, temperature, and dewpoint forecasts to avoid losses due to crops rotting in the field and the additional costs of artificially drying hay and grains; precipitation forecasts to minimize losses in drying raisins; and humidity forecasts to reduce costs and losses resulting from poor conditions for drying tobacco.

These requirements have been identified in studies by Rapp and Huschke (1964), Dancer and Tibbitts (1973), Theron *et al.* (1973), Bussell *et al.* (1978), and PROFS (1981). Other sources of information on agricultural requirements for very-short-range forecasts include a paper by Mason (1966), a WMO report (World Meteorological Organization, 1968), the proceedings of a conference on the value of meteorological services (Royal Meteorological Society, Australian Branch, 1979), and a technical note by Omar (1980). Some of these studies also have contained estimates of the economic values of current and/or improved very-short-range weather forecasts for agriculture. These benefits include $74 million per year for improved nowcasts in the United States (Dancer and Tibbitts, 1973); R12.3 million for improved forecasts in South Africa (Theron *et al.*, 1973); and £6.5 million for improved short-range forecasts in the UK (Bussell *et al.*, 1978).

2.2 CONSTRUCTION

The major construction industry requirements for very-short-range forecasts can be summarized as follows: (a) forecasts of precipitation timing, duration, and intensity and of wind speed to avoid damage to finished work (e.g. concrete) and to minimize the costs of protecting exposed surfaces, structures, and work sites; and (b) forecasts of precipitation, wind speed, and high and low temperatures to schedule work in an efficient manner. Specific information concerning these requirements is contained in several studies, including Rapp and Huschke (1964), Russo (1966), and Bussell *et al.* (1978). The paper by Russo (1966) also presented estimates of the benefits to the United States construction industry that would result from better use of current weather information – $0.5 to $1.0 billion per year – and from perfect short-range (i.e., 0–24 h) forecasts – an additional $0.3 billion per year. Other estimates of the economic values of nowcasts to the construction industry have been made by Mason (1966), Bhattacharyya *et al.* (1975), and Bussell *et al.* (1978).

2.3 ENERGY

Very-short-range forecasts of temperature, humidity, wind speed and direction, cloud cover, and other parameters are required by the electric power and gas industries in order to optimize procedures related to the generation and distribution of electricity and natural gas. Moreover, forecasts of thunderstorms, strong winds, low temperatures, and freezing precipitation are needed by electric power companies to minimize damage to lines and equipment and to schedule repairs. These requirements are described in works by Rapp and Huschke (1964), Mason (1966), WMO (1968), and Bhattacharyya *et al.* (1975). Estimates of the annual economic value of such forecasts to the electric power industries in France (F33 million), West Germany ($9 million), and Australia (A$0.1 million) are included in a WMO report (WMO, 1968); an estimated annual value for the UK (£0.2 million) is given by Mason (1966); and an estimate of the annual value of improved forecasts for the United States electric power industry ($36 million) is provided by Bhattacharyya *et al.* (1975).

2.4 TRANSPORTATION

Airlines and air terminal companies need very-short-range forecasts of ceiling height and visibility, strong winds and turbulence, and surface ice and snow conditions in order to minimize risks and maximize efficiency in pre-flight and in-flight decisions and to make effective adjustments to weather-related fluctuations in traffic. In addition, optimal route planning by airlines necessitates forecasts of wind speed and direction, as well as of severe weather and icing conditions, along the flight path. Similarly, truckers, motorists, and public transportation systems require forecasts of snowfall, precipitation amounts, and other storm-related events in order to avoid damage to weather-sensitive goods, select optimum routes, prevent accidents, minimize delays, and maximize revenues under conditions of adverse weather. These requirements have been identified by Rapp and Huschke (1964), Mason (1966), WMO (1968), Bhattacharyya *et al.* (1974), Bussell *et al.* (1978), and PROFS (1981). Substantial economic values for nowcasts have been estimated for the airline industry in these studies, with considerably smaller benefits derived for other segments of the transportation industry.

2.5 PUBLIC SAFETY AND GENERAL PUBLIC

Organizations responsible for public safety include public health groups (e.g. hospitals, clinics, doctors), disaster control agencies, and highway maintenance and traffic control organizations. The very-short-range forecast requirements of these groups include (a) forecasts of temperature and humidity extremes, as well as forecasts of significant changes in these quantities, to alert hospitals, clinics, and the public to weather conditions that may seriously aggravate certain health-related illnesses; (b) forecasts directly or indirectly related to potentially dangerous or damaging natural events – for example, tornado and severe thunderstorm forecasts, severe wind forecasts, forecasts related to storm surges and avalanches, and precipitation and stream-flow forecasts in the case of floods – to minimize loss of life and property damage; and (c) forecasts of snowstorms, surface icing, visibility, and other events (e.g. floods) to enable highway maintenance and traffic control organizations to take appropriate actions to reduce risks of traffic accidents and protect roads from damage. The general public requires very-short-range forecasts of rain, snow, wind, and temperature to assist in planning activities such as commuting, recreation, and shopping.

The above-mentioned requirements have been identified by Rapp and Huschke (1964), Bergen and Murphy (1978), Bussell *et al.* (1978), and PROFS (1979, 1981). Relatively few reliable estimates of the values of very-short-range forecasts to the general public are available. However, efforts to estimate the value of nowcasts in the area of public safety have suggested that significant benefits ($105 million per year) would be realized by United States snow removal agencies with improved snow forecasts (Bhattacharyya *et al.*, 1975) and that the value of very-short-range forecasts in reducing the damage caused by floods and other natural disasters is significant as well.

2.6 SUMMARY

Some currently available information concerning user requirements and economic values of very-short-range weather forecasts is summarized in Table 1. The user-requirement information in Table 1, based on the requirements discussed in the previous subsections, consists of data related to the activity or hazard, the weather elements or events of interest, and the lead time of the forecasts. Events and lead times are not indicated in cases where they could not be inferred from the studies in question.

The information in Table 1 relating to economic values of very-short-range forecasts includes the type of analysis (i.e. method of value assessment) used in the study and the estimated value of the forecasts. These value estimates generally are not directly comparable due to differences in the approaches, methodologies, assumptions, and currencies used in the various studies. Moreover, the data in Table 1 should be considered to be a selective rather than a comprehensive description of the requirements and values for very-short-range forecasts.

3 Deficiencies in current data and need for further studies

Even a cursory analysis of previous studies reveals that current data regarding user needs are deficient in several important respects. The deficiencies relate to both the dimensions of user requirements and the methods used to obtain the relevant data, and they can be described in terms of the availability (or completeness) and reliability of such data. The nature of the various dimensions and methods, as well as the deficiencies in the current state of knowledge, are examined briefly in subsections 3.1 and 3.2, respectively. Subsection 3.3 contains a short discussion of the need for further studies of user requirements for nowcasts, including some specific suggestions regarding possible approaches and methods.

3.1 DIMENSIONS OF USER REQUIREMENTS

User requirements for very-short-range forecasts can be described in terms of a number of characteristics or dimensions of such information: (a) variable or event; (b) spatial domain; (c) temporal domain; (d) lead time; (e) form of the forecasts; (f) content and format of information packages; and (g) communication or dissemination media. In order to define user requirements adequately,

TABLE 1

Summary of selected user-requirement and economic-value data

Activity/ hazard	Reference	Region	Event(s)[a]	Forecast lead time (hours)	Analysis type(s)[b]	Economic values (millions/year)
Agriculture						
Daily planning	Mason (1966)	UK			BOE	£20
Scheduling	Bhattacharyya *et al.* (1975)	US	TS		DM,BOE	$232
Crop spraying	WMO (1968)	W. Germany			BOE	$3.5
	Dancer and Tibbitts (1973)	US	P,W	0–12	S,BOE	$44.6
	Bussell *et al.* (1978)	UK	P	1–6	BOE	£1.0
Harvesting	Dancer and Tibbitts (1973)	US	P,T,DP	0–12	BOE	$2.8
	Bussell *et al.* (1978)	UK	P	1–6	BOE	£3.5
Planting	Dancer and Tibbitts (1973)	US	P,W	0–12	BOE	$3.1
Frost/heat Protection	Dancer and Tibbitts (1973)	US	T,W,C,DP	0–12	BOE	$23.8
All	Theron *et al.* (1973)	S. Africa	H,R,Fr,Su,T,DP,W	0–48	S	R12.3
	Dancer and Tibbitts (1973)	US	P,T,W,C,DP	0–12	BOE	$74.3
	PROFS (1979)	US[c]	R,Sn,H,W,T	0–12	S	$1.8
Construction						
All	Russo (1966)	US	P,T	0–24	BOE	$500–1,000
	Bhattacharyya *et al.* (1975)	US	TS		DM	$922
	Bussell *et al.* (1978)	UK	P	1–6	BOE	£10
Energy						
Electric power production	Mason (1966)	UK	T,DP,C		BOE	£0.2
	WMO (1968)	France	T,P		BOE	F33
	WMO (1968)	W. Germany	T		BOE	$9
	WMO (1968)	Australia	R,T		BOE	Aus. $0.1
	Bhattacharyya *et al.* (1975)	US	T,TS		EM,BOE	$36
Gas distribution	Mason (1966)	UK	T		BOE	£0.2
	WMO (1968)	France	T		BOE	F34
	Bhattacharyya *et al.* (1975)	US	T		EM	$3.3

Transportation						
Highway accidents	Bussell *et al.* (1978)	UK	P	1–6	BOE	£0.5
Total aviation	Bhattacharyya *et al.* (1975)	US	TS	2	DM,BOE	$30
Airlines	Mason (1966)	UK	W,St		BOE	£6.5
	WMO (1968)	Australia	St		BOE	Aus. $14
	WMO (1968)	France	W,St		BOE	F36
	WMO (1968)	W. Germany	W		BOE	$1.8
(crash risk)	Bussell *et al.* (1978)	UK	TS	1–6	BOE	£1
(delays)	PROFS (1979)	US[d]	St	0–12	BOE	$24.4
Public safety and general public						
Snow removal	Bhattacharyya *et al.* (1975)	US	Sn		DM	$105
Flood	Bhattacharyya *et al.* (1975)	US	St		BOE	$29
	Bussell *et al.* (1978)	UK	P	1–6	S,BOE	£3.8
Hurricane	Bhattacharyya *et al.* (1975)	US	HL		BOE	$2–3
Forest fire	Bhattacharyya *et al.* (1975)	US	TS		BOE	$40
Wind storm	Bergen and Murphy (1978)	Boulder, US	W	2–4	S	$0.2
Recreation	PROFS (1979)	Denver, US	R, Sn, H, W, T	3	S	$7.3
(boating)	Bhattacharyya *et al.* (1975)	US	St		BOE	$4.6
Commuting	PROFS (1979)	Denver, US	R,Sn,H,W,T	0.5	S	$13.3
Shopping	PROFS (1979)	Denver, US	R,Sn,H,W,T	0.5	S	$10.4

[a] C = Cloud cover; DP = Dew point; Fr = Frost; H = Hail; HL = Hurricane landfall; P = Precipitation; R = Rain; Sn = Snow; St = Storm; Su = Sunshine; T = Temperature; TS = Thunderstorm; W = Wind

[b] BOE = Back-of-the-envelope calculation; DM = Decision model; EM = Economic model; S = Survey.

[c] Region within 100 km of Denver. [d] Chicago, Denver, Los Angeles, and New York airports.

detailed and reliable information must be obtained concerning all these dimensions. First, the nature of the variable or event itself must be specified. For example, should the meteorological element of interest to the user be treated as a continuous variable or as a set of two or more categories or ranges of values? With regard to spatial domain, should the information of concern be applicable to a point or an area? What are the horizontal dimensions of the area? The issue of temporal domain relates to whether the requirements are for essentially instantaneous values, for averages or other statistics over an interval of time, or simply for occurrences and non-occurrences of events in specific time periods. Lead time, obviously, is an important dimension of user requirements. How much advance warning do users require in order to make effective use of the forecasts in their particular activities or operations?

The form that the forecasts should take is an important but frequently neglected characteristic of user requirements. This dimension involves several factors including the mode of presentation of the information (words in a text, numbers in a table, a figure or diagram), the use of numbers and/or words, and the mode of expression of the uncertainty inherent in the forecasts. Since most forecasts contain information related to several variables or events, issues such as the content and format of such forecast packages also are important dimensions of user requirements. For example, how much information should be included in such packages (the problem of "information overload")? How should the information be arranged or ordered to ensure its efficient and effective use? Finally, the communication and dissemination media required to provide the information in a convenient and timely manner must be given careful consideration in any study of user requirements. It should be noted that the dimensions of user requirements are interrelated, and studies of these requirements should recognize that needs in one dimension may have important implications for needs in other dimensions.

Previous studies of user requirements, such as the works summarized in section 2, have provided useful information concerning some dimensions of these needs in certain situations. However, very few studies have contained a *complete* description of requirements in terms of all the above-mentioned dimensions. Obviously, information concerning each dimension is required to provide an adequate basis for designing and implementing forecasting systems to meet user needs. Moreover, to realize this objective, the information concerning requirements must be reliable (or credible) as well as complete. Reliability refers to the extent to which the information accurately reflects the true needs of users. This aspect of user-requirement data depends largely on the methods used to determine these requirements, and this topic will be addressed briefly in the next subsection.

3.2 METHODS USED TO DETERMINE REQUIREMENTS

Two general approaches have been taken to the problem of determining user requirements for weather forecasts. These two approaches are referred to here as the *passive* and *active* approaches. In the passive approach, data related to user requirements are obtained primarily by evaluating the requests for forecasts received from actual and potential users and by analysing the feedback provided by current users concerning the usefulness and appropriateness of the forecasts. Such an approach may provide valuable information regarding the nature of user requirements in certain areas of application or for some specific decision-making situations, but it generally yields an incomplete and insufficiently detailed set of data on which to base the design of a forecasting system or even the tailoring of specific forecast products.

The active approach to determining user requirements involves conducting surveys of and/or interviews with the relevant users. It is relatively easy to prepare a questionnaire or interview schedule and to conduct a study based on such "instruments". However, the quality and utility of the data obtained from such an investigation are directly related to the insight and care with which the study is designed and conducted. General questions regarding the needs or preferences of users frequently will elicit answers either lacking in specificity or identifying requirements the dimensions and scope of which greatly exceed the true needs of the users. Since many users will not be familiar with the types of information that could be provided or, in some cases, even with the way in which the information should be used, it is necessary for the individuals conducting such studies to be well acquainted with the information-sensitive activities of the users and to ask specific questions that users can readily relate to their particular decision-making situations.

Surveys and interviews that attempt to include a wide spectrum of users of forecasts necessarily must treat user requirements in a superficial manner or must be inordinately long and time-consuming, thereby taxing the patience and attention span of the participants. Thus, the scope of such studies should be relatively narrow, with emphasis on obtaining detailed information concerning user requirements in a particular decision-making situation or in a small set of similar or related situations. In summary, the design, conduct, and analysis of the results of proper user-requirement studies involving surveys and interviews are difficult tasks demanding great care and attention to detail, as well as knowledge of the users' operations. Moreover, some familiarity and experience with questionnaire design and survey sampling procedures are also invaluable. Unfortunately, very few studies of user requirements conducted to date have satisfied these conditions.

The passive and active approaches to the problem of determining user requirements both suffer from an even more basic defect in that they rely on users' *perceptions* of their needs. That is, such requirement data are *not* based on either prescriptive or descriptive models of the information-processing and decision-making procedures of individual users. The potential role of such models in further studies of user requirements is discussed in the next subsection.

3.3 NEED FOR FURTHER STUDIES

As indicated above, the information currently available concerning user requirements for very-short-range weather forecasts suffers from several basic deficiencies. In particular, this information is based primarily on perceptions of users' needs as reflected by their responses to hypothetical questions in surveys and interviews. Moreover, studies of user requirements involving such procedures generally devote little if any attention to the values or benefits of forecasts, thereby ignoring the fundamental relationship between user requirements and economic values. What is needed is an approach to determining user requirements that takes this relationship into account and formulates prescriptive and/or descriptive models of users' information-processing and decision-making procedures. We believe that *decision analysis* provides such an approach.

Decision analysis is a procedure for analysing and modelling complex, uncertain decisions in a logical and rational manner (Raiffa, 1968; LaValle, 1978). A decision–analytic approach to the problems of determining user requirements and estimating economic values of forecasts appears to possess several attractive features. First, decision analysis focuses on individual decision makers or users, so that user requirements can be defined directly in terms of the basic elements of actual decision-making problems (e.g. actions, events, probabilities of events, utilities of consequences). Moreover, since this approach generally involves the formulation of decision-making models, quantitative user-requirement data must be obtained in sufficient detail to support the modelling effort. Second, the concept of the value of information plays a key role in decision analysis. In this framework, information is considered to be of value only if it changes the user's decision. Moreover, the values of different types of information can be assessed and compared using a decision–analytic approach. Third, an intimate relationship between user requirements and economic values is readily apparent in this approach, since requirements for information are meaningful only if the information can be demonstrated to be of value to the user. This situation suggests, as previously indicated, that studies involving the analysis and modelling of decision-making problems will provide the most useful and credible data concerning both user requirements and economic values of weather forecasts. Surveys and interviews can be used to obtain important input to the modelling effort within the overall decision–analytic framework.

Relatively few studies of user requirements and economic values of weather forecasts have been based on decision-making models. The familiar cost-loss ratio model (e.g. Thompson, 1962; Murphy, 1977) – and extensions of this model –

have been used to study the use and value of forecasts in several decision-making situations (e.g. see Kolb and Rapp, 1962; Hashemi and Decker, 1969; Thompson, 1972; Kernan, 1975). However, these models are greatly over-simplified representations of the decision-making problems of concern and generally have employed an *ex post* approach to these problems rather than the *ex ante* approach that is consistent with decision analysis. Only recently have studies of requirements and/or values of forecasts been based on an *ex ante* decision–analytic framework. Work of this type includes the investigations by Baquet *et al.* (1976) and Katz *et al.* (1982) concerning the value and use of frost forecasts by orchardists and the analysis by Howe and Cochrane (1976) regarding the benefits of short-range forecasts of snowstorms for urban snow removal. These studies have provided valuable insights into the use of forecasts by specific groups of decision makers and have yielded useful data concerning both requirements and values for the situations of concern. Further investigations of this type in the future will be required to obtain more detailed and credible information concerning user requirements and economic values of very-short-range weather forecasts.

4 Summary and conclusion

We have summarized briefly some data currently available concerning user requirements for very-short-range forecasts. This summary suggests that requirements for such forecasts do exist among user groups in a variety of sectors and that significant economic values are associated with these needs. However, the current data regarding requirements and values are deficient in several important respects, and these deficiencies raise serious questions concerning the completeness and reliability of such data. Surveys and interviews are inherently flawed as methods of determining requirements and values (although they may be useful when employed in conjunction with other, more suitable procedures). Decision analysis was described as an approach that can provide reliable data concerning requirements and values, and the principal attributes of this approach were briefly discussed. For the interested reader, an introduction to decision analysis with a meteorological flavour (Winkler and Murphy, 1982) will appear in a forthcoming book. We believe that decision analysis, and other methods based on efforts to model users' information-processing and decision-making procedures, are the most promising means of determining user needs and economic values for very-short-range forecasts.

It is encouraging to note that recent programs concerned with system design, development, and implementation of improved observing, forecasting, and dissemination procedures have given increased attention to the problems of determining user needs and economic values. Studies conducted in conjunction with the PROFS (Prototype Regional Observing and Forecasting Service) program in the United States (Beran and Little, 1979) and the FRONTIERS (Forecasting Rain Optimized using New Techniques of Interactively Enhanced Radar and Satellite) program in the UK (Browning, 1979) provide some good examples of such work. Only through continued efforts of this type involving careful analysis of user requirements and economic values – and employing the best available methodology – can the meteorological and user communities realize the full benefits of these programs.

References

Baquet, A. E., Halter, A. N. and Conklin, F. S. (1976). The value of frost forecasting: a Bayesian appraisal. *Am. J. agric. Econ.* **58**, 511–520.

Beran, D. W. and Little, C. G. (1979). Prototype regional observing and forecasting service from concept to implementation. *Nat. Weath. Dig.* **4**, 2–5.

Bergen, W. R. and Murphy, A. H. (1978). Potential economic and social value of short-range forecasts of Boulder windstorms. *Bull. Am. met. Soc.* **59**, 29–44.

Bhattacharyya, R. K., Greenberg, J. S., Lowe, D. S. and Sattinger, I. J. (1974). Some economic benefits of a synchronous earth observatory satellite. Princeton, New Jersey, ECON, Inc., Final Report (NASA Contract NAS 5-20021), 138 pp.

Bhattacharyya, R. K., Steele, W., Greenberg, J. S., Stevenson, P. and Fawkes, G. (1975). Economic benefits of STORMSAT: an initial survey. Princeton, New Jersey, ECON, Inc., Final Report (NASA Contract NAS 5-20021), 181 pp.

Browning, K. A. (1979). The FRONTIERS plan: a strategy for using radar and satellite imagery for very-short-range precipitation forecasting. *Met. Mag.* **108**, 161–184.

Bussell, R. B., Cole, J. A. and Collier, C.G. (1978). The potential benefit from a national network of precipitation radars and short period storm forecasting. London, England, Ministry of Agriculture, Fisheries and Food (MAFF), Paper, 47 pp.

Dancer, W. S. and Tibbitts, T. W. (1973). Impact of nowcasting on the production and processing of agricultural crops. *In* "Multidisciplinary Studies of the Social, Economic, and Political Impacts Resulting from Recent Advances in Satellite Meteorology", vol. 5. University of Wisconsin, Space Science and Engineering Center, Madison, Wisconsin. Report (NASA Grant NGL 50-002-114), pp. 1–47.

Hashemi, F. and Decker, W. (1969). Using climatic information and weather forecast for decisions in economizing irrigation water. *Agric. Met.* **6**, 245–257.

Howe, C. W. and Cochrane, H. C. (1976). A decision model for adjusting to natural hazard events with application to urban snow storms. *Rev. Econ. Statist.* **58**, 50–58.

Katz, R. W., Murphy, A. H. and Winkler, R. L. (1982). Assessing the value of frost forecasts to orchardists: a dynamic decision-making approach. *J. appl. Met.* **21**, 518–531.

Kernan, G. L. (1975). The cost-loss decision model and air pollution forecasting. *J. appl. Met.* **14**, 8–16.
Kolb, L. L. and Rapp, R. R. (1962). The utility of weather forecasts to the raisin industry. *J. appl. Met.* **1**, 8–12.
LaValle, I. H. (1978). "Fundamentals of Decision Analysis". Holt, Rinehart and Winston, New York.
Mason, B. J. (1966). The role of meteorology in the national economy. *Weather* **21**, 382–393.
Murphy, A. H. (1977). The value of climatological, categorical and probabilistic forecasts in the cost-loss ratio situation. *Mon. Weath. Rev.* **105**, 803–816.
Omar, M. H. (1980). The economic value of agrometeorological information and advice. Geneva, Switzerland, World Meteorological Organization, Technical Note No. 164, 52 pp.
PROFS (1979). Report of a study to estimate economic and convenience benefits of improved local weather forecasts. Boulder, Colorado, NOAA, Environmental Research Laboratories, NOAA Technical Memorandum ERL PROFS-1, 18 pp.
PROFS (1981). User service requirements document. Boulder, Colorado, NOAA Environmental Research Laboratories, Report, four sections plus appendix.
Raiffa, H. (1968). "Decision Analysis: Introductory Lectures on Choices under Uncertainty". Addison-Wesley, Reading, Massachusetts.
Rapp, R. R. and Huschke, R. E. (1964). Weather information: its uses, actual and potential. RAND Corporation, Santa Monica, California. Memorandum RM-4083-USWB, 126 pp.
Royal Meteorological Society, Australian Branch (1979). "Proceedings of the Conference on Value of Meteorological Services". Melbourne, Australia, 21–23 February, 1979, 197 pp.
Russo, J. A. (1966). The economic impact of weather on the construction industry of the United States. *Bull. Am. met. Soc.* **47**, 967–972.
Theron, M. J., Matthews, V. L. and Neethling, P. J. (1973). The economic importance of the weather and weather services to the South African agricultural sector – a Delphi survey. Pretoria, South Africa, Council for Scientific and Industrial Research, Information and Research Services, Techno-Economics Division, Research Report 321, 134 pp.
Thompson, J. C. (1962). Economic gains from scientific advances and operational improvements in meteorological prediction. *J. appl. Met.* **1**, 13–17.
Thompson, J. C. (1972). The potential economic benefits of improvements in weather forecasting. San Jose, California, California State University, Department of Meteorology, Final Report (NASA Grant NGR 05-046-005), 80 pp.
Winkler, R. L. and Murphy, A. H. (1982). Decision analysis. *In* "Probability, Statistics, and Decision making in the Atmospheric Sciences" (A. H. Murphy and R. W. Katz, Eds). Westview Press, Boulder, Colorado. In preparation.
World Meteorological Organization (1968). The economic benefits of national meteorological services. Geneva, Switzerland. World Meteorological Organization, World Weather Watch Planning Report No. 27, 55 pp.

1.2
Designing a Very-short-range Forecasting System

DONALD W. BERAN and ALEXANDER E. MACDONALD

1 Introduction

In this chapter we first draw attention to the various elements that make up a total very-short-range forecasting system. Then we describe how an accepted design philosophy is being newly applied to this complex problem.

A very-short-range forecast system can be defined by contrasting its features with those of a mid- to long-range system (Table 1). Examination reveals significant differences in all components.

2 Observations

The data base for synoptic systems rests primarily on the 12-hourly radiosonde upper air data and hourly surface observations, supplemented by satellite images of cloud position, type, and motion. Radar contributes little to forecasts having lead times greater than 6 h. The data set as it now exists is usually inadequate input to very-short-range forecasts because the time and space resolution for upper-air and surface data is not sufficient for following changes that occur in less than a few hours.

Despite its negligible role in synoptic-scale work, radar is the cornerstone of our present efforts to provide very-short-range services because of its ability to detect and track severe and other important weather conditions that have life cycles of less than 6 h. The added capability of Doppler radar ensures that these ground-based weather detection devices will continue to play the central role in very-short-range forecasts. Satellite data provide an effective bridge from the larger mesoscales and forecast periods of several hours to the convective storm scale and warning times of minutes. While radar will certainly remain as a key nowcast tool, it is the combination of radar and satellite data that promises a significant increase in our ability to deal with the mesoscale (Browning and Collier, 1982).

TABLE 1
Comparison of forecast systems

Characteristic	Mid-to long-range	Very-short-range
Forecast lead-time	>6 h	0–12 h
Meteorological scale	Synoptic	Mesoscale
Area of coverage	Global or continental	Local or regional
Nature of forecast	General	Site-specific
Observations		
Time interval	Many hours	Fraction of hours
Spacing	Thousands of km	Tens of km
Data volume	Medium $\sim 10^6$ bits per h	Large $\sim 10^8$ bits per h
Data flow	Slow (minutes to hours)	Rapid (seconds to minutes)
Forecast method	Numerical, statistical	Single-cell or system extrapolation (now-casting), mesoscale models, statistical
Dissemination	Slow, passive	Rapid, both active and passive

The qualitative nature of satellite and radar information has been a barrier to its more extensive use in existing operations. The subjective assessments that are derived from satellite and radar images are valuable (Purdom, 1982), but leave untapped the benefits that would be possible if quantitative information were available from these sensors. The research community has already demonstrated that wind fields can be directly measured with Doppler radar (Wilson and Wilk, 1982) and that the thermal structure of the atmosphere can be sensed from satellite (Kelly *et al.*, 1982). Ground-based systems that combine radar and radiometric elements to measure vertical profiles of wind, temperature, and humidity have also been demonstrated (Little, 1982). A fundamental challenge for the designers of advanced very-short-range forecasting systems will be to determine the optimum combinations of these new data sets with existing information.

3 Forecasts

Synoptic forecasting depends heavily upon numerical and statistical methods. Forecasters use automated analyses and guidance information provided by large centralized computers and numerical models. We are not likely soon to see a total extrapolation of these capabilities to the mesoscale. While rapid, automated processing and analysis are essential as the space–time resolutions of the data base are increased, neither the mesoscale models nor the computer capacity to drive them appear imminent. Nor do the appropriate observational data yet exist to initialize them. Mesoscale forecasting will depend on collection, processing, and display of dense data sets (for forecasting) and the use of extrapolation and simple models that quantify a particular physical process (such as orographic forcing of precipitation).

The nature of very-short-range forecasts will also depend on the available data base, amplifying the importance of proper selection, integration, and design of the mesoscale observing system. Indeed, an optimum very-short-range forecast service will result only from a careful balance between the needs of the forecaster and the characteristics of the data base. Such a design is still difficult to achieve because few complete mesoscale data bases exist for use by the developer of mesoscale forecast techniques.

New forecast techniques are, quite naturally, closely tied to the data base available to the researcher. Thus, we have the growth of radar meteorology and satellite meteorology as separate and distinct fields "blessed" with their professional societies and service organizations alike. These separate research groups have developed workable forecast techniques and no doubt will continue to do so. However, specific sensor-driven research leaves unanswered a basic question: Would some combination of data sets produce more effective techniques? If the observational requirements for the mesoscale are determined by input from each of the now separate observation communities, we are likely to find that the operational systems would be prohibitively expensive and complex. On the mesoscale, especially, we must begin to think more of the forecaster's requirements for a combined data set that may rely only on portions of the data from each sensor system. Until we can accomplish these integrations and the determination of forecasting requirements, we are in danger of designing a system that must make all possible measurements, almost regardless of their worth, simply because we cannot effectively judge what is, in fact, required.

4 Dissemination

When we are capable of producing sufficiently accurate, site-specific, 0–12 h forecasts, we must also be able to deliver this information to just those people who need it to make a timely decision. Our present methods of disseminating weather information cannot meet this challenge.

Dissemination can occur in two modes: passive or active. Passive dissemination includes all methods whereby the information is simply transmitted or published for a general audience. If the information is wanted, it is available, but there is no special attempt to be overly specific or to get the user's attention. Examples of passive dissemination are NOAA Weather Radio and the weather forecast of the evening TV news. Active dissemination, on the other hand, includes all those methods that make direct contact with a user because he has a specific need to know. Examples are radio contact that warns a pilot of wind shear and a telephone call to someone in the path of a tornado.

The distinction between active and passive becomes sharper as we consider the design of a very-short-range service. Passive dissemination is adequate for synoptic-scale forecasts simply because there is plenty of time for the user to obtain his required amount of weather information according to his own schedule. Very-short-range information is, of course, highly perishable – a 1-h forecast is not much good 6 h later. The smaller-scale of very-short-range events (an area of thunderstorms or a tornado) makes it of interest to a much smaller group of people. Hence, the information must be given promptly to only those who will be affected.

When such warnings cover an overly broad area, we run the danger of losing credibility. Everyone who is warned about a severe storm, but who is not affected by the storm, wonders about the credibility of the forecaster; they generally will not blame the dissemination system.

Critical policy issues relating to who has the authority and responsibility to warn people of impending danger must also be examined. The information provided by improved techniques will stimulate rapid and, at times, costly reactions by the user. There will be no margin of tolerance for the confusion that could result if different warnings were issued from separate sources.

New technology and a close working relationship between government and the private sector will be essential in dissemination. The practice in the United States of giving government forecasts to the commercial media for dissemination will continue, but we must also begin to develop more direct forecaster-to-user links to ensure that the user receives important information with an absolute minimum delay. Such links might use automated forecast products in a pre-arranged format. We must consider technology that would automatically ring the telephone or activate the radio or TV of just those people who are in danger. Also essential information should be selectively accessible on the initiative of a user. Such methods, when coupled with computer-generated voice response and a carefully designed emergency service alerting network, will be essential components of future very-short-range forecasting services.

5 Communication

The communication network is the thread that binds any forecast system together. We have already seen the significant transition from teletype and facsimile to the National Weather Service AFOS (Klein, 1978) system in the United States. The absolute requirement for rapid reaction as forecast lead times are reduced will place ever greater demands on the communication system.

This, and the potentially greater data volumes and rates, make it essential to design even more efficient communication systems to support very-short-range forecast systems.

Optimized data base management, data compression, and efficient command languages will all be necessary. Simply transmitting large streams of raw data from sensors will not be an acceptable mode. Automated methods that extract only significant information will help to reduce the load on the communication network. New developments in communications bode well for future operational systems. Satellite distribution of weather information and fibre optics technology are only two examples of the many exciting possibilities.

6 A design method

The preceding sections illustrate the large and complex problem of designing a very-short-range forecasting system. The designers must balance the cost and effectiveness of each of the elements (radar, satellite, forecast mode, voice response, etc.) in each of the components (observing, communications, etc.) to achieve an optimum design that will serve a broad spectrum of users. Classic system design methodology, whereby the needs of the ultimate user determine the dissemination, forecast, and data requirements, can be a useful starting point. Yet the assumptions that must be made, and the fact that weather forecasting is an art as well as a science, will frustrate the pure system designer.

To address this problem we must use the following design methodology. Once the design concept is developed we must ask which of the vast array of possible elements will answer the fundamental questions: What is really needed to make a good forecast? and what is a good forecast? Using the excellence of a forecast as the measure of design quality, we can evaluate the elements of the forecast system as variables that may or may not contribute. Because human judgment alone may be biased or fallible, we must also draw upon statistics for our evaluation. This method could be applied to existing synoptic systems that have been in operation for a long period. Forecast output prior to the introduction of a new device or technique could be compared with the output that followed. The length of record for this kind of performance examination would need to be very long to ensure that a statistically significant number of each type of weather event has been forecast by many different meteorologists. Since no very-short-range mesoscale-oriented system now exists, we obviously cannot use this historical method.

However, we can adapt this approach to provide solid evidence of the worth of various system elements in a relatively short period of time. First, we must select a difficult kind of forecast that is of importance to the user; for example, predicting the affected area and lead time for severe storms. Second, we should assume that if a system is designed to produce this forecast, then it can effectively handle a broad set of less challenging situations. Third, we should create a quasi-operational forecast office containing as many as possible data sets, workstation configurations, and other necessary system elements. Fourth, a complete data base for many events (both severe and fair weather) must be collected. Finally, these data are to be played back to a large number of forecasters while one or more of the key elements is either provided or not provided until a statistically significant sample is created. The impact of a given data set or forecast technique can then be examined in terms of the size of the warning area and forecast lead time.

This black box approach – where the output of the box (the forecast system under test) is affected by various inputs – will provide significant information for the system designer. The record and playback procedure also eliminates the wait for significant events to occur naturally. If we were to rely on real-time operation only, it would be necessary to operate a given configuration for several years before enough storms occurred to provide similar answers, and the storm-to-storm variability would increase the level of doubt in the final answer.

7 Prototype regional observing and forecasting service (PROFS)

The above systems design approach is central to PROFS, a program for producing optimum functional design specifications for operational forecast systems. If it is successful, the result will be well-tested performance specifications for systems that demonstratively improve very-short-range weather services. Such information is essential to the system designer and weather service manager alike.

The first tests of this design and element selection method have begun. Data collected during the 1981 Colorado severe storm season by satellite, radar, a mesoscale surface network, and conventional methods have been stored for later playback. Three forecaster work stations, which range from AFOS to an advanced interactive processing and display unit, are ready for testing with forecasters from National Weather Service operational offices, the USAF Air Weather Service, other NOAA components, the private sector, and universities. The results from this test will be available by early 1982 and will be used by the PROFS System Design group to modify preliminary design concepts. By mid-1982, the modified system will be put together in the PROFS Exploratory Development Facility and a similar but more refined series of tests will take place. The design that evolves from this second iteration will be demonstrated in 1983 and will provide functional and performance specifications for an improved very-short-range forecast service system.

References

Browning, K. A. and Collier, C. G. (1982). This volume, pp. 47–61.
Kelly, G. A. M., Forgan, B. W., Powers, P. E. and LeMarshall, J. (1982). This volume, pp. 107–121.
Klein, W. N. (1978). Introduction to the AFOS program. Proceedings of Conference on weather forecasting and analysis and aviation meteorology, pp. 186–193. American Meteorological Society, Boston, Massachusetts.
Little, C. G. (1982). This volume, pp. 65–85.
Purdom, J. F. W. (1982). This volume, pp. 149–166.
Wilson, J. W. and Wilk, K. E. (1982). This volume, pp. 87–105.

1.3

Blueprint for the Future Swedish Weather Service System

SVANTE BODIN

1 Introduction

Our rapidly changing society and the development of new technologies have called for a new look at the weather service system – from observations to dissemination of weather information. The present system is essentially built on out of date technology and philosophy. The additions and developments which have taken place after the Second World War have not been fully integrated in the present-day weather service. Many countries face a similar situation to Sweden. New needs and demands from society simply cannot be met with the system we have today.

At the Swedish Meteorological and Hydrological Institute this situation has been a painful experience which eventually has led to something of great importance – a design for a completely new concept of the future service incorporating a large nowcasting component. This chapter presents the major ingredients and philosophy of the new program, known as PROMIS-90–Program for an Operational Meteorological Information System. Although requiring investment it will give ample returns in the form of improved services to society. But perhaps most of all it provides a possibility of meeting the challenges of the future. A full documentation of the PROMIS-90 system can be found in Bodin *et al.* (1979).

2 Premises and assumptions

The design of the weather service must stem directly from the information requirements. As a starting point weather information requirements have been carefully analysed. This analysis has been based on sources both within and outside Sweden. Among them can be mentioned a comprehensive investigation of the Swedish construction industry by Haag (1978) and enquiries made by the

Swedish Radio and Television company. The results – concerning both the general public and industry and their need for both warnings and general weather information – point to large and increasing demands for very-short-range, detailed forecasts as well as longer forecasts – 5 days to a month – which the present-day weather service cannot satisfy. It also seems clear that large profits can be made by rationally using weather forecasts in many areas, e.g. construction, energy production and snow clearing. Typical cost-benefit ratios are about 1/100. The future weather service must be designed so that these demands can be met.

The development of the weather service must also be viewed against the following technological and scientific developments.

i. A survey of developments and research within meteorology shows that computers will become even more important in forecasting, taking over more and more of the steps in producing 1–10 day forecasts. The basis for this computerisation is the advanced dynamical/mathematical models with subsequent statistical interpretation that have been developed and recently made operational. To capitalize on these computer products a new weather service system requires only fairly small investments. As a member of ECMWF (European Centre for Medium range Weather Forecasts) Sweden can utilize the latest developments in numerical modelling in operational forecasting on the time scale 2–10 days. The development of a national limited area model for 12–36-h forecasts will provide the basis for the forecast interval between very-short-range forecasts and ECMWF-products.

ii. The development in the computer field is very favourable to investments in computer hardware. Computer components are expected to fall in price by between 25–45% in the 4 years up to 1985. On the other hand expenses for programming and system development will grow.

iii. A modern weather service already uses large volumes of data. A transition to digital transmission techniques will lead to considerably higher communication speeds. An analysis shows, however, that costs would become fairly high if the existing Swedish data network were to be used for transmitting radar and satellite pictures, each of which contains about one megabyte of information. An optimization can be achieved if decentralized computer facilities are used to translate alphanumerical information to graphical output media instead of sending pictures in digital form. An alternative is to use the new satellite Tele-X for broad band transmissions.

iv. Remote sensing techniques for observing various atmospheric parameters are of great importance for the future. In this context one should mention the emergence of satellite systems giving pictures with high horizontal resolution in several spectral ranges, for example TIROS-N and METEOSAT, see, e.g. Liljas (1982). Surface-based microwave radiometers and Doppler radar systems might eventually replace the old type of radiosonde stations. The automatic station technology is essentially already fully developed and very useful for providing detailed observations in a dense net.

v. Modern weather radar systems are capable of quantitatively measuring precipitation rates. The information can be transmitted digitally to the user and be subject to further treatment in the form of colour TV displays (Browning and Collier, 1982). Information of this kind has proved to be of great value for very-short-range forecasts and warnings (Austin and Bellon, 1982).

vi. Graphic radar and satellite information can also be directly disseminated to individual customers.

Parts of the technical advances listed above, including graphical display and dissemination systems, have already been integrated in operational weather service systems; see, e.g. Leep (1981).

3 System analysis

The design of an integrated weather service system for both short-range and long-range forecasts was preceded by a careful analysis of requirements for observations, analysis and prediction methods, and final products disseminated to the users of weather information. This analysis has been carried out for different time and space scales and suggests that one should make a division between forecasts for the range 0–12 h ahead and from 12 h to 10 days ahead. This conclusion is based on the real-time requirement in nowcasting and very-short-range forecasting and on the observational and forecast methods used to produce the forecast. In order to keep down data transmission to a minimum and to allow fast responses between the forecaster and the customers the very-short-range forecasts are more conveniently made and disseminated by Regional Weather offices (RW). Forecasts for periods longer than 12 h ahead, based on central computer models, are more efficiently produced at a Central Weather office (CW). The result of the analysis is summarized in Table 1, which shows for different forecast periods and associated space scales, the required observations, data processing and prediction methods as well as products to be disseminated.

The nowcasting at the Regional Weather office (RW) requires real-time observations, transmission of data and presentation to the meteorologist. This can only be accomplished by an automated system. On the other hand the forecast methods depend very much on the judgment of the meteorologist. In the very-short-range forecasting system a relatively dense automatic station network, weather radars and satellite information are used.

The Central Weather office (CW) is less demanding concerning manual work. Forecast production for the range 12–36 h ahead will largely be based on

TABLE 1

Overview of requirements for observations, data processing and dissemination for different time and space scales. Observations are discussed further in the general text. LAM means Limited Area Model, a regional numerical forecast model of synoptic scale flow. RW means Regional Weather office

Time and space scales	Observations
0–2 h 0–100 km "Nowcasting"	Complete regional radar coverage. Continuous operation Automatic stations (including buoys). Regional network for wind, temperature, humidity with spacing ~ 40 km. Wind measurements in narrow navigation channels with spacing less than 20 km. Wind and temperature along popular mountain tracks. Temperature, wind, humidity and radiation along highway sections prone to slipperiness. All values in real time 1–2 vertical remote sounding systems for wind, temperature and humidity. Measurements every hour Reports from civilian and military aircraft in the region (ASDAR, AIREP, QBC) Airport observations, hourly synoptic observations and METAR In south Sweden METEOSAT digital information every half hour
2–6 h 20–300 km	Complete radar coverage Complete synoptic observations every third hour, spacing 80 km Automatic stations (including buoys). Pressure measurements with spacing ~ 50 km, wind, temperature and humidity with spacing ~ 40 km once an hour Digital (TIROS-N) satellite pictures with a period of 3–6 hours 1–2 vertical sounding systems at least every 6th hour Scandinavian synoptic observations every third hour Acoustic sounders, instrumented masts etc.

Data processing	Products for dissemination
Fast presentation of data, automatic monitoring of weather, alarm function Manual interpolation of data and analyses of aircraft reports. Automatic analyses of wind, temperature and radar echoes Manual linear and dynamical/ physical extrapolation Automatic linear and statistical extrapolation, advection models, convection models, radiation models and statistical forecasts Manual and automatic monitoring of forecasts Verification	Selective warnings for high wind speed in coastal areas, lakes, mountains and at exposed inland locations, for thunderstorms, hail, rainstorms, freezing precipitation, heavy snow, slipperiness (public and special customers) Detailed information about actual weather, precipitation, temperature and wind (on request, by, e.g. telephone) Detailed forecasts of weather, precipitation, temperature and wind (when needed by public and customers) Aviation TRENDS and SIGMETS
Quality control and presentation of data, automatic monitoring and alarm Manual analyses of weather maps. Automatic analyses of pressure, pressure tendency, satellite pictures and local boundary layer parameters Manual linear and dynamical/ physical extrapolation. Automatic extrapolation of pressure field. Physical prediction of fog. Radiation computations. Local boundary layer model. Statistical predictions Manual and automatic monitoring of forecasts Archiving and verification	Warnings as above (decreased selectivity in space) Detailed forecasts in time and space of precipitation, temperature and weather Detailed forecasts to special customers Terminal Area Forecasts (TAF), VFR forecasts, SIGMETS

Continued overleaf

Time and space scales	Observations
6–18 h 20–300 km	Synoptic observations, every third hour. Spacing ~ 80 km Automatic stations with pressure sensor every third hour. Spacing ~ 50 km Digital satellite pictures with period 3–6 hours (TIROS-N) Satellite vertical soundings (e.g. TOVS) every 6th hour or more frequently 1–2 vertical sounding systems every 6th hour Foreign observations (SYNOP, TEMP, PILOT, AIREP) every 3rd or 6th hour Ship observations Acoustic sounders, masts etc.
12–36 h 150–4000 km	As above
30 h–10 d 300–10 000 km	As above

Data processing	Products for dissemination
Control and presentation of data. Automatic monitoring Manual analyses of weather map. Automatic analyses of pressure, pressure tendency, weather map, upper air charts, boundary layer parameters every 3rd and 6th hour (centrally) Manual dynamical/physical extrapolation (RW). Automatic extrapolation/advection of pressure field, LAM, boundary layer model, statistical interpretation (of LAM). Later mesoscale model (CW) Manual and automatic monitoring of forecasts Archiving and verification	General warnings for gales and storms in coastal areas, heavy precipitation, slipperiness and frost in parts of region, and risk of thunderstorms and forest fire Forecasts to the public for different districts of the region Forecasts to ships in coastal areas of wind, visibility, icing and ice conditions Forecasts to special customers. Probability forecasts for different places of precipitation, minimum/maximum temperature, frost, critical wind speeds, humidity, cloudiness, thunderstorms, air pollution level, pollen content, sea water level etc. according to demand by customers TAF, upper winds and temperatures
Control and presentation of data. Automatic monitoring. Inversion of satellite radiance data to temperature Manual completion of automatic analyses of surface weather map. Automatic analyses of upper air data and satellite information. Analyses of boundary layer LAM, Boundary Layer Model, statistical interpretation, manual inspection and modification – new interpretation Manual monitoring of forecasts Verification and archiving	General warnings of gales and storms, road slipperiness and other traffic hazards, forest fire risks Forecasts to general public for regions or parts of regions Forecasts to ships of wind, visibility, weather, icing and ice conditions in the seas surrounding Sweden Forecasts to special customers. Probability forecasts and other forecasts according to demand by customers
Statistical interpretation of forecasts from foreign weather institutes (e.g. Germany, USA, England and ECMWF) Manual editing of automatic text Producing dissemination material Verification and archiving	Forecasts to general public. General forecasts for the second and third day. Average temperature and accumulated precipitation for day 4–10. Probable weather situations. Forecasts for holiday resorts Forecasts to special customers, e.g. probability of temperature and precipitation for day 2–5. Accumulated precipitation for day 1–5 and 6–10. Interpretation for specific places.

numerical and statistical forecasts produced by the central computer. For the range + 36 h to + 10 days the forecasts of the ECMWF will be utilized with subsequent regional statistical interpretation.

4 Dissemination to external users

Greater emphasis will be put on efficient dissemination of weather information. New media like Text TV and the Swedish video text system, Datavision, will be used. The new weather service system permits forecasts and weather information to be designed to meet the specific needs of customers. A large number of customers, individuals, organizations and commercial enterprises will obtain forecasts tailored automatically according to predefined criteria. Increased contacts between meteorologists and users will eventually develop a new awareness of possibilities for using weather information efficiently.

5 System design

As part of the PROMIS-90 study, we have gone into some detail in the design of the technical system and the interactions between man and machine. The system design is divided into basically two parts – the regional systems and the central system. The regional systems are designed to meet the real-time, nowcasting, requirements. At the core is a telecommunication system to carry out the transmission of large data volumes from the local observation sources as well as from/to the central weather office. The system is conceived in a modular fashion to ensure high system reliability.

The entire system is based on six Regional Weather offices (RW) and one Central Weather office (CW) in Norrköping. The regional offices may be located in Stockholm, Malmö, Göteborg, Sundsvall, Norrköping and Luleå.

5.1 REGIONAL WEATHER OFFICES (RW)

A regional weather office will be built up around the basic systems for observations, data communication, dissemination and data display. Each *regional* observation network (Fig. 1) will consist of about:

- fifty automatic stations with pressure, temperature, wind and humidity sensors (300 altogether in Sweden);
- fifteen hybrid synoptic stations (totalling 90);
- one-to-two aerological stations/vertical sounding systems (10–20 altogether);
- two-to-three weather radars (totalling 16);

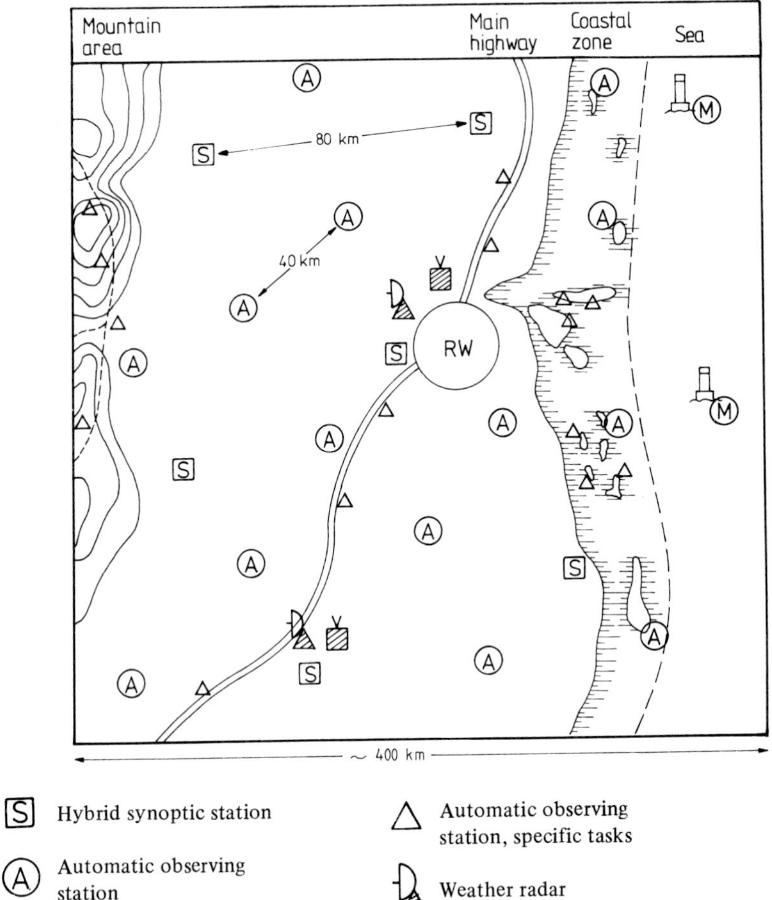

Hybrid synoptic station

Automatic observing station

Maritime automatic observing station

Automatic observing station, specific tasks

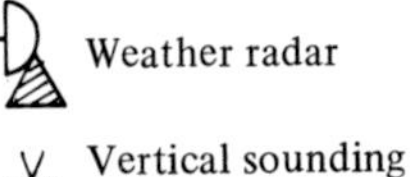
Weather radar

Vertical sounding station

Fig. 1. Typical deployment of observing stations in a region (not all stations are shown).

- high-resolution, TIROS-N, satellite information and METEOSAT data from CW;
- auxiliary information – acoustic sounders, mast data, additional automatic stations in specific regions, e.g. mountain tracks and major highways, observations from aircraft etc.

The hybrid synoptic station consists of an automatic station observing wind, temperature, humidity and pressure and a human observer who monitors the remaining synoptic weather elements. The latter are collected every third hour while the automatic part can be tapped every ten minutes.

Radiosondes will probably be used at least for another 10 years to obtain temperature and humidity observations from the free atmosphere. However, a fairly rapidly advancing development of ground-based microwave radiometers for vertical profiling of temperature and humidity in the troposphere is taking place (Little, 1982). Together with satellite-based microwave radiometers a new sounding system might be developed within the next ten-year period. PROMIS-90 foresees this development and is ready to make full use of such a facility when it becomes available. Such a system would mean a substantial increase in the resolution in both time and space of the observations of the troposphere and lower stratosphere. The new systems will eventually replace the old radiosondes.

Concerning the weather radars it will be possible for each regional office to obtain digital radar data from adjacent regions to monitor weather systems crossing regional boundaries.

The technical system of the Regional Weather office is designed in a modular fashion. It consists of a basic data base comprising 8 Mbytes and various display media such as

1 graphical CRT
4 alphanumerical displays
1 electrostatic plotter
2 colour TV displays
1 printer

The system structure is shown in Fig. 2.

5.2 CENTRAL WEATHER OFFICE (CW)

The design of the central system is based on a RW located in Norrköping. It presupposes an updated computer system which in 1986 should have a capacity of at least 10 MIPS (millions of instructions per second). The system comprises an additional real time data base of 15 Mbytes and:

1 electrostatic plotter
4 alphanumerical displays
3 colour TV displays
1 digital scanner, to make it possible to introduce manually produced charts into the system.

The whole design also comprises system maintenance, service and back-up facilities.

6 Development plan

The development of the new weather service system is just about to start. At present a detailed specification has been worked out to form the basis for

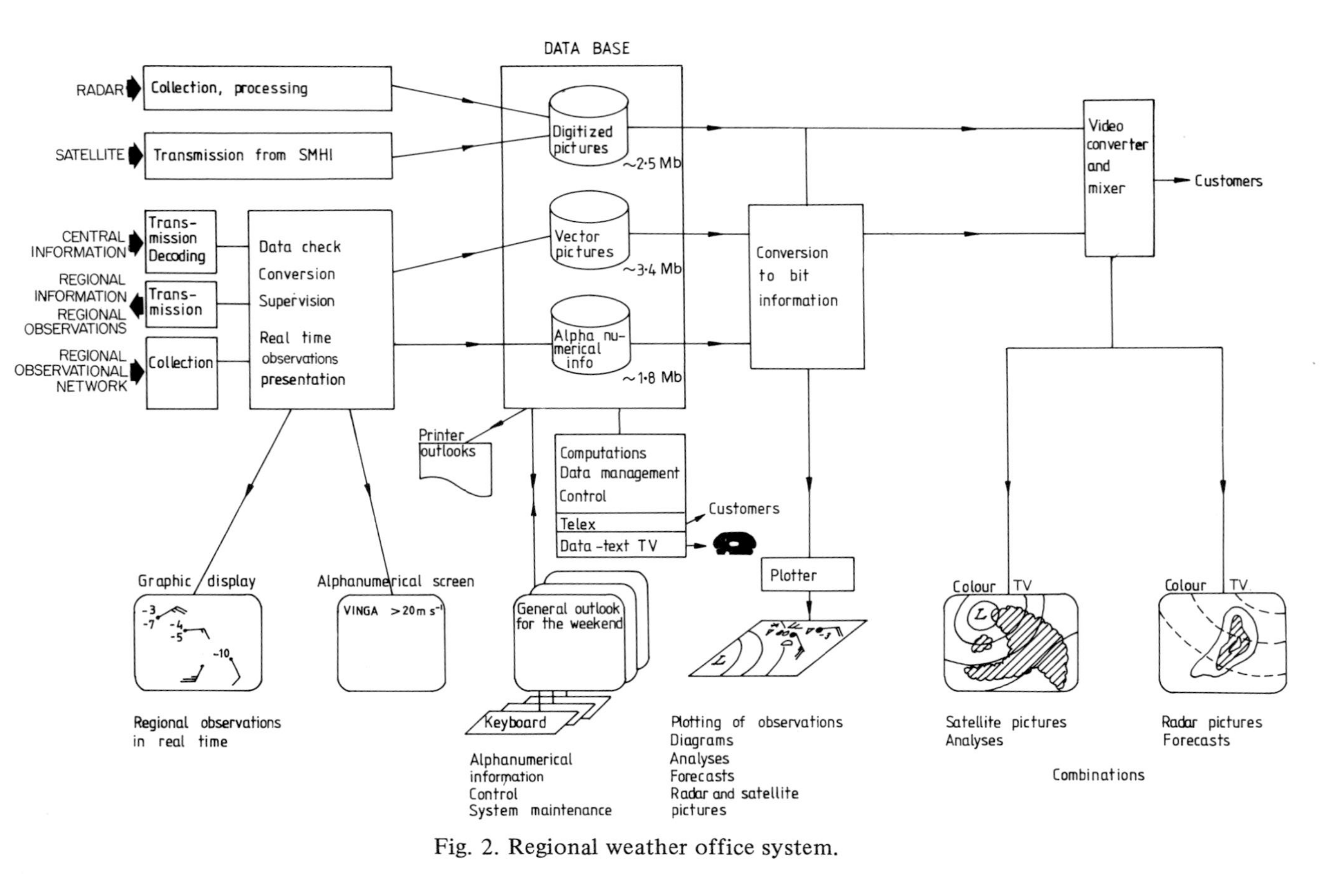

Fig. 2. Regional weather office system.

purchase of equipment for the first experimental Regional Weather office which is going to be built up in Norrköping, starting in 1983. The experimental RW will change smoothly into a prototype for a Regional Weather office, including facilities for local dissemination of forecasts.

The development of the Central Weather office depends on the upgrading of the central computer system, which is planned to take place at the latest in 1986 when all numerical prediction models as well as statistical and dynamical interpretation schemes will be ready for operational use. In fact the new 9-level, primitive equation, limited area model has recently been put into operational use and a three-dimensional boundary layer model is just about ready for initial tests on the CRAY-computer of ECMWF. This means that the Central Weather office can be operational in 1987 and implementation of RWs can then take place. However, the PROMIS-system is composed of independent modules, such as the satellite imagery system, that can be taken into operation when they are ready.

The experimental RW-office will also be used for testing the sub-systems. The testing of a new weather radar system with Doppler facilities will be included. An essential part is also to decide the optimum mix between automatic stations and synoptic stations. The environmental conditions for people working with the new equipment must also be investigated at an early stage.

References

Austin, G. L. and Bellon, A. (1982). This volume, pp. 177–190.

Bodin, S. Liljas, E. and Moen, L. (1979). The future weather service at the Swedish Meteorological and Hydrological Institute – PROMIS 90. SMHI. (In Swedish.)

Browning, K. A. and Collier, C. G. (1982). This volume, pp. 47–61.

Haag, T. (1978). Weather dependence of the construction industry. SMHI Report, No. RMK 11 (1978). (In Swedish.)

Leep R. (1981). Weathervision – a forecasting and dissemination tool for the 1980s. *In* "Nowcasting: Mesoscale observations and short-range predictions", pp. 299–302. Proceedings of a symposium at the IAMAP General Assembly, 25–28 August 1981, Hamburg. European Space Agency, ESA SP-165.

Liljas, E. (1982). This volume, pp. 167–176.

Little, C. G. (1982). This volume, pp. 65–85.

1.4
The Mesoscale Observational Network in Japan

RYOZO TATEHIRA, MICHIO HITSUMA and YOSHIHISA MAKINO

1 Introduction

In recent years it has been recognized that the way to improve local weather forecasting lies in understanding mesoscale weather systems. The Japanese Islands have been almost completely covered by a weather radar network since 1971. To supplement the weather radar, an automated mesoscale surface observational network, called AMeDAS, has been in operational use since 1979. The Japanese Geostationary Meteorological Satellite, providing visible and infrared images, was launched in 1977 and has been in operational use since 1978. The above three systems are the major tools for observing mesoscale weather in Japan. The features of these systems are briefly described in this chapter.

The spatial resolution of the mesoscale observational networks is:

1. Weather radar network (20 sites) – resolution of the order of 2 km.
2. AMeDAS (Automated Meteorological Data Acquisition System) – 1317 stations to monitor amount of precipitation (average spacing of 17 km), and 838 stations to observe surface wind, temperature and sunshine (average spacing of 21 km).
3. GMS (Geostationary Meteorological Satellite System) – a visible channel with resolution of about 2 km in the Japanese Islands, and an infrared channel with resolution of about 7 km in the Japanese Islands.

Mesoscale systems can be monitored in detail by the combined use of these mesoscale networks. Very-short-range prediction is possible by extrapolating the observed mesoscale systems.

Heavy precipitation is of great importance in Japan, since it often causes serious damage or even disaster. In general, heavy precipitation is directly associated with mesoscale systems and often shows remarkably local concentration. Therefore, the primary purpose of the mesoscale networks in Japan is the forecasting of precipitation.

2 The radar network

Since the volume of data from the radar network is very large, computers are used as far as possible to achieve rapid processing. High-speed processing and dissemination are indispensable for very-short-range forecasting, because of its short lead time and valid time. In order to process the initially analogue radar data by computer, first the precipitation echo must be separated from the coexisting ground clutter, then it must be digitized with a suitable mesh size, say about 5 km.

A technique of ground clutter rejection has been developed recently and implemented on several radars in Japan (see Fig. 1). This technique is based on the fact that the mean power of precipitation echo is approximately proportional to the variance of the amplitude of linear-detected signal, irrespective of the intensity of ground clutter in the same echoing volume (Tatehira and Shimizu, 1978). The suppression ratio of ground clutter achieved by filtering the variance is limited to about 35 dB due mainly to the small and slow (but significant) fluctuation of ground clutter intensity. A method for achieving further improvement without degrading the accuracy of intensity measurement of precipitation echo superimposed on the ground clutter has been described by Tatehira and Shimizu (1980).

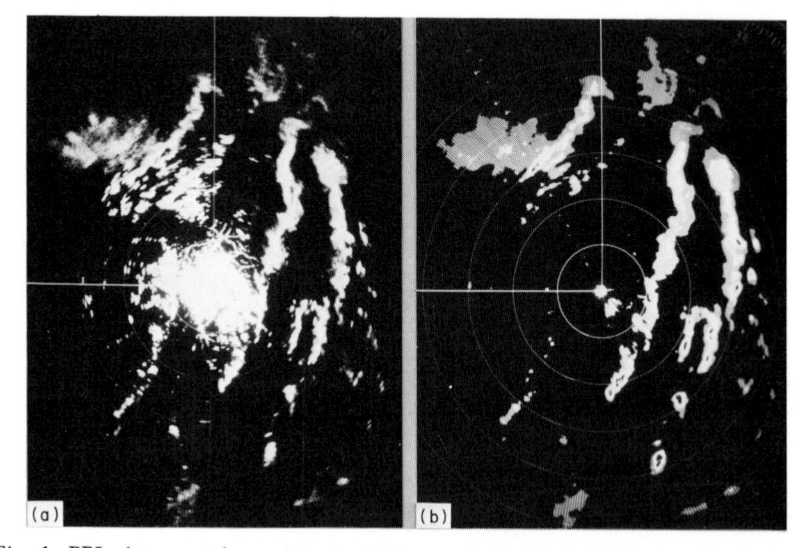

Fig. 1. PPI pictures taken with the Tokyo radar at 1548 GMT on 4 September 1979 (elevation: 0.0°; range marks: every 50 km). (a) Precipitation echoes together with ground clutter on a conventional display. (b) After ground clutter rejection (grey: rainfall intensity of 1–4 mm h^{-1}; white: 4–16 mm h^{-1}; black: more than 16 mm h^{-1}).

Two weather radars (Nagoya and Fukui) in central Japan were equipped with a digitizer in 1982. The main features of the digitizer are as follows:

1. suppression of ground clutter;
2. digitization of echo patterns from PPI scans at 3 elevations (*c.* 1°, 2°, 3°) eight times an hour, and construction of an approximately constant-altitude (CAPPI) echo pattern from them;
3. integration of the above CAPPI echo patterns over periods of 1 h;
4. digitization of echo patterns from PPI scans at 10 elevations once every hour and calculation of echo top height;
5. hourly transmission of the echo data (2), (3) and (4) on a 5-km grid to the high-speed computer at the Japan Meteorological Agency in Tokyo;
6. transmission of the echo data (2) to the colour graphics display at the appropriate regional forecast centre (and to users through public telephone lines if the need arises).

The Japan Meteorological Agency is planning eventually to install the digitizer on all the weather radars in Japan (the Mt Fuji and Niigata radars in central Japan are to be equipped at the end of 1983). Composite echo patterns will be derived on a 5-km grid, and transmitted to the forecast centres by high-speed facsimile (4800 b.p.s. Coded Digital). An example of a composite echo pattern covering the whole of the Japanese Islands is shown in Fig. 2, but for a few years to come the routine composite will be derived from only four radars in central Japan (i.e. Nagoya, Fukui, Mt Fuji and Niigata). The composite echo pattern will be used for the calculation of very-short-range forecasts by computer.

3 The automated meteorological data acquisition system (AMeDAS)

The AMeDAS is designed to collect weather data obtained using unmanned instruments located at about 1300 observing stations throughout Japan, and to send them to about 60 forecast centres, NHK (Japanese Broadcasting Corporation) and the Japan Weather Association (the largest meteorological consulting corporation in Japan). Major features of this system are:

1. observation, collection and distribution of data are fully automated;
2. the data distribution is completed within 20 minutes after the observation time;
3. the data collection is made over the telephone switching network under computer control;
4. the system is operated on a 24-h, year-round basis.

There are four types of observing stations in the AMeDAS, as seen in Fig. 3.

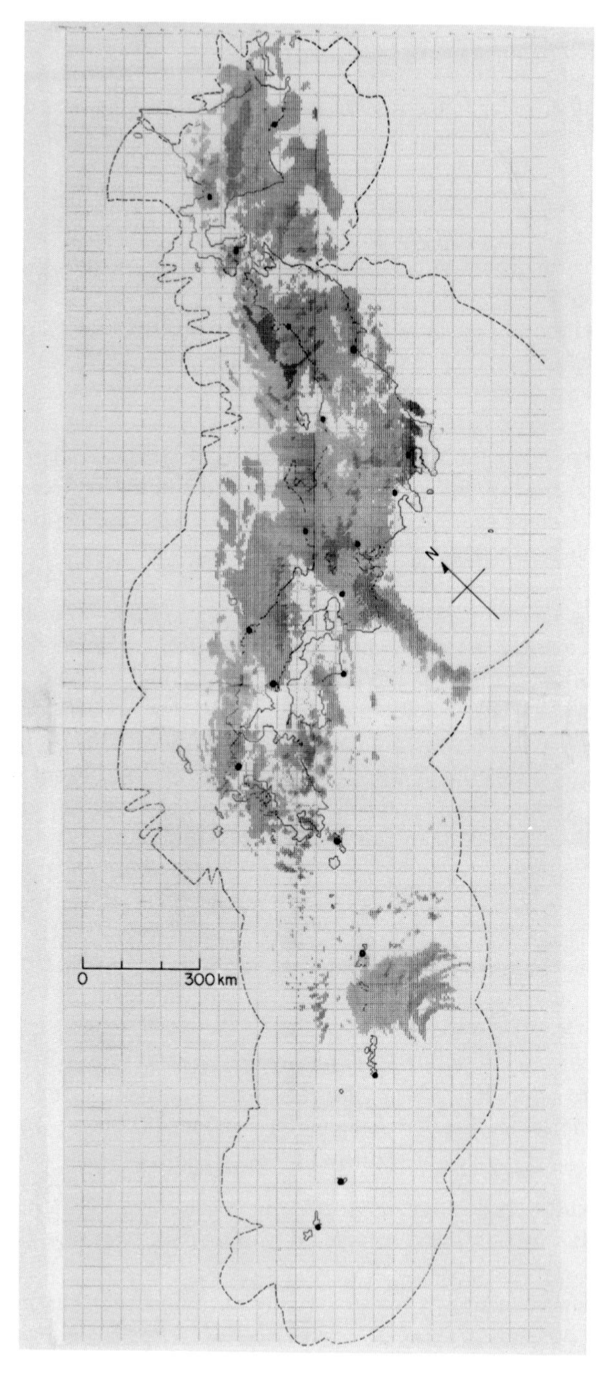

Fig. 2. Composite echo pattern at 00 GMT 9 September 1976 (Typhoon 7617 Fran). The echo patterns at 3 intensity levels from 20 radar sites (dots) in Japan were digitized on a 5-km grid and composited by computer. Broken lines enclosing the composite echo pattern show the range at which the lowest beam is at a height of 4 km.

1. automatic observing equipment installed in meteorological observatories: elements observed are wind direction and speed, temperature, precipitation amount and sunshine duration (151 stations);
2. automatic weather stations at unmanned sites: elements observed are the same as in (1) (687 stations);
3. automatic precipitation stations at unmanned sites (203 stations);
4. automatic precipitation stations at unmanned sites in mountainous regions: observed data are transmitted by radio to the responsible meteorological observatory (276 stations).

The data from all these stations are sent to the AMeDAS computer centre in Tokyo through the public telephone lines. The computer centre automatically calls all observing stations every hour and collects data from each. The collected data are processed in the form of maps and tables, and then sent to the terminal equipment of every regional forecast centre and local forecast centre (see Fig. 3). In addition, the latest data observed every ten minutes are sent in response to requests through the terminal equipment. Dedicated lines operating at 1200 b.p.s. are used to connect the computer centre with the regional forecast centres; lines operating at 200 b.p.s. connect with the local forecast centres. Access to the data of any station for any observation time within the past 24 hours can be obtained through the terminal equipment by the use of a special call code.

Figure 4a is the chart of AMeDAS wind data and hourly precipitation (mm) in a Typhoon situation. Figure 4b shows wind vectors on a 10-km grid. These have been interpolated by computer from the surrounding AMeDAS wind data, taking into consideration topographic effects in a simple way. That is, the grid values are calculated by averaging the surrounding AMeDAS wind data with a weight which is related not only to the distance but also to the topography between the grid and the respective AMeDAS station. Various features of wind field (divergence, vorticity, etc.) are easily calculated from these grid values. We are developing techniques to use these features to forecast the future development of radar echo in order to improve on the accuracy of very-short-range prediction achieved by simple extrapolation.

4 The geostationary meteorological satellite system

The Geostationary Meteorological Satellite (GMS) of Japan is part of the global satellite system of five geostationary satellites. GMS is also used as part of the mesoscale observational network. The output data of GMS are classified into two categories: various types of cloud picture (on-line processing) and various products derived from cloud images (batch processing). We use the cloud pictures, but the derived products are obtained too infrequently to be useful for monitoring mesoscale systems.

45°N
145°E
140°E
40°N
35°N
140°E
135°E
35°N
0
100
200 km
130°E
30°N

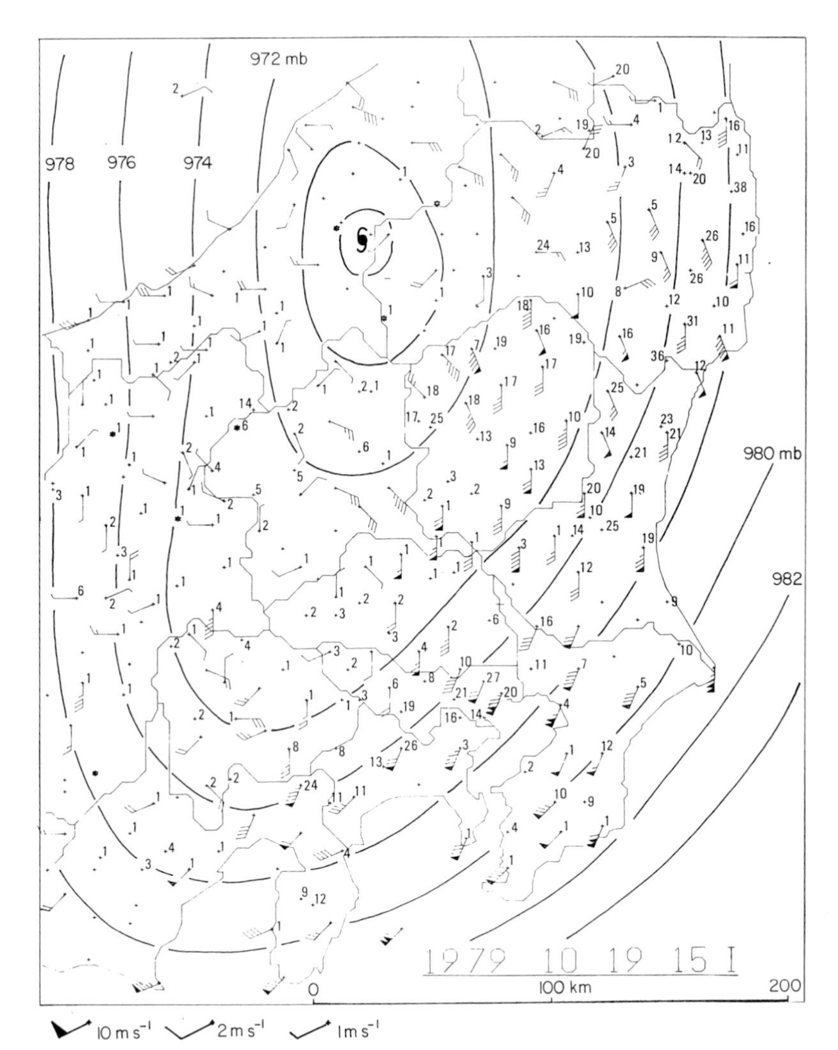

Fig. 4a. Surface winds and hourly precipitation amounts given by the AMeDAS at 06 GMT 19 October 1979 (Typhoon 7920 Tip).

Fig. 3 (*opposite*). Observing stations in the AMeDAS.

●, ■, □, Automatic observing equipment installed in regional forecast centres, local forecast centres, and meteorological observatories, respectively.

○, Automatic weather stations at unmanned sites.

△, Automatic precipitation stations at unmanned sites.

∧, Automatic precipitation stations (radio) in mountainous regions.

Since regional forecast centres (●) need the AMeDAS data from a wider area than the local forecast centres (■), regional forecast centres are equipped with terminal equipment of 1200 b.p.s. compared with 200 b.p.s. at local forecast centres. The 31 stations in the Ryukyu Islands are not shown in this figure.

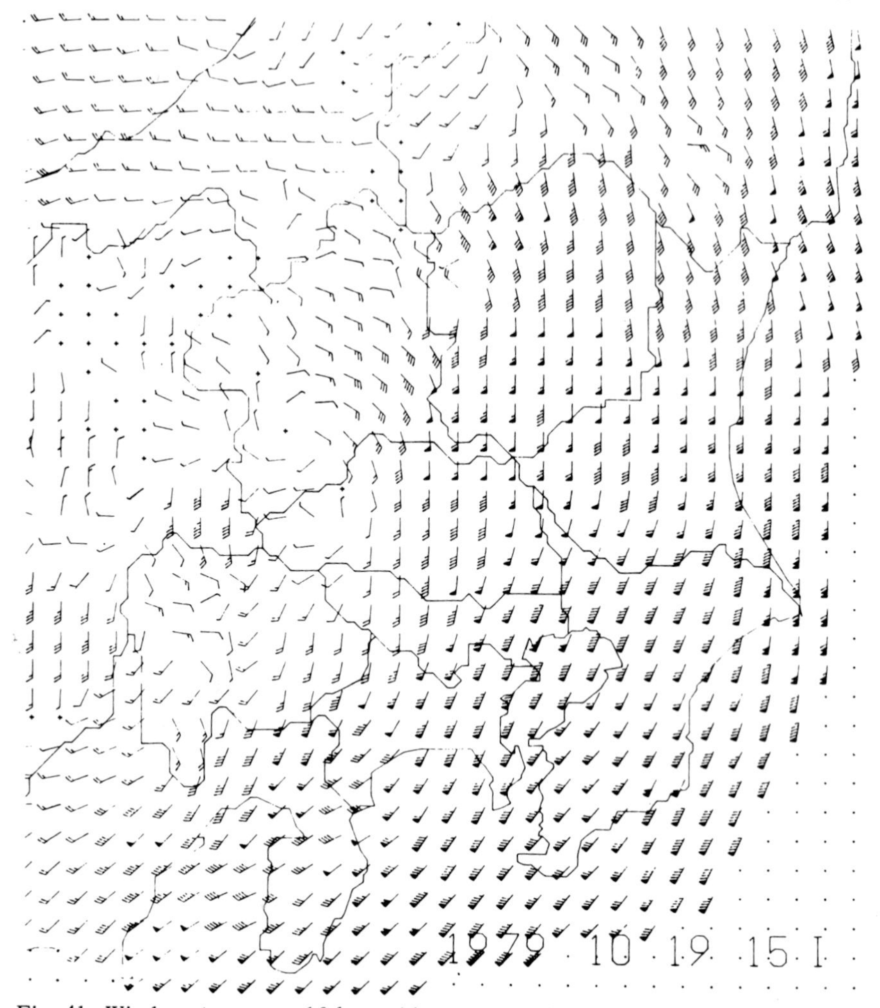

Fig. 4b. Wind vectors on a 10-km grid, estimated from the AMeDAS wind data obtained in Fig. 4a.

It is intended that the time delay between the initial observation time and dissemination of very-short-range forecasts shall be between 15 and 30 min. Unfortunately at present the GMS cloud pictures are not received until about 40 min after real time. Thus the radar network data and the AMeDAS data play the dominant part in very-short-range forecasting of precipitation in Japan, because they are received almost in real time. The effective range of weather radar is, however, limited to 300 km at most and so satellite images with wider coverage will be needed to extend the lead time of very-short-range forecasts.

Furthermore, the satellite images offer the possibility of detecting mesoscale systems before they develop into precipitation-producing systems. They can also identify systems which will intensify the pre-existing precipitation system by interaction. We are currently carrying out tests to combine the satellite picture with the radar echo pattern and the AMeDAS data on a colour graphics display using the man–computer interactive method of Browning (1981).

References

Browning, K. A. (1981). A total system approach to a weather radar network. *In* "Nowcasting: mesoscale observations and short-range prediction", pp. 115–122. Proceedings of a symposium at the IAMAP General Assembly, 25–28 August 1981, Hamburg. European Space Agency, ESA SP-165.

Tatehira, R. and Shimizu, T. (1978). Intensity measurement of precipitation echo superposed on ground clutter – a new automatic technique for ground clutter rejection. Preprints, 18th Radar Meteorology Conference, pp. 364–369. American Meteorological Society, Boston, Massachusetts.

Tatehira, R. and Shimizu, T. (1980). Improvement in performance of ground clutter rejection. Preprints, 19th Radar Meteorology Conference, pp. 176–179. American Meteorological Society, Boston, Massachusetts.

1.5

An Integrated Radar-satellite Nowcasting System in the UK

K. A. BROWNING and C. G. COLLIER

1 Introduction

The past decade has seen technological progress in three important areas: digital processing and transmission of radar rainfall data; frequent cloud imagery from geostationary satellites; development of man–computer interactive displays. These advances have permitted the development of nowcasting systems by means of which it will be possible to derive in real time reliable analyses of the distribution of precipitation and, also, very-short-range forecasts by linear extrapolation. In this chapter we describe such a system being developed in the UK. Our intention is to emphasize the factors that have influenced the system design rather than to present a mere technical description. Although the system which we discuss focuses on just precipitation, a good analysis of precipitation patterns contributes to the understanding of the general weather situation and therefore leads to improved analysis and forecasting of other weather variables too.

2 The total system

The overall system design is shown in Fig. 1. It is an all-digital system based on a distributed network of computers (PDP 11 series). Components of the system are:

a. *multiple radar sites*, with a mini-computer at each site to preprocess the data into a convenient form for transmission, display and further manipulation – the newer radars are unmanned so as to have low operating costs and so that they may be sited in remote areas (e.g. hill tops) to provide unobstructed coverage;

b. *a central network computer* which automatically combines the data from the radars to provide integrated radar coverage over a large area;

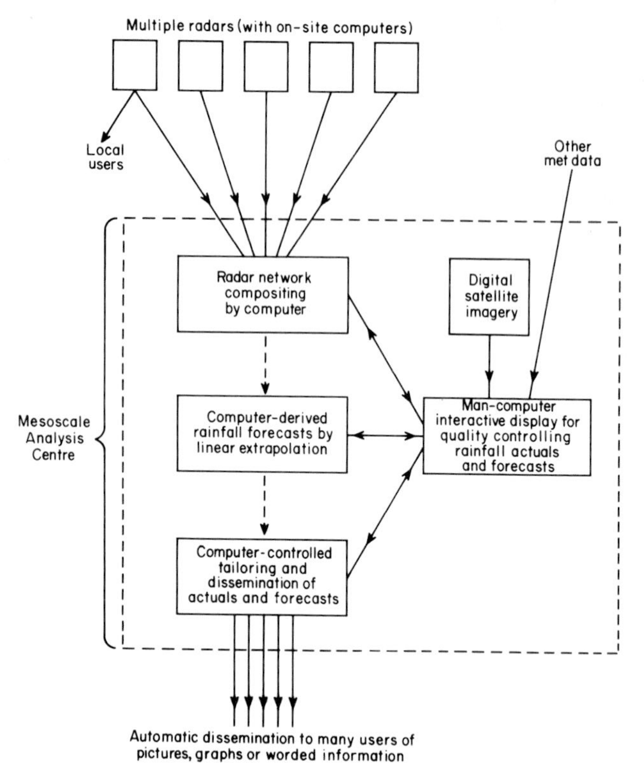

Fig. 1. The total system design.

c. *a satellite ground receiver* (Primary Data User Station) with a computer to preprocess the cloud imagery from the geostationary satellite, METEOSAT;
d. *computer-derived forecasts* using linear extrapolation methods;
e. *a man–computer interactive image display* that enables a centrally located meteorologist to compare radar, satellite and other data, and to carry out analyses of the combined data set;
f. *automated tailoring and dissemination* of actual and forecast precipitation information in user-orientated formats – special-purpose receivers are being used at present (section 3) but more versatile media such as viewdata (PRESTEL) are likely to be used in future.

Items (b) to (f) are co-located and together they constitute a Mesoscale Analysis Centre (Browning, 1980); (b) and (c) use separate mini-computers (PDP 11 series), whilst (d), (e) and (f) share a midi-computer (VAX 11/750).

We have been operating items (a) and (b) routinely since 1979, together with a limited dissemination of precipitation actuals (Collier, 1980). The other items are in various stages of development. Item (c) was established in 1981 with the launch of METEOSAT 2. Item (d) exists in rudimentary form. The most novel aspect of the system is the interactive display (item (e)) and work now underway is expected to lead to the implementation of a working system in 1983.

Precipitation actuals and very-short-range forecasts derived from them are highly perishable products and it is crucial to get them to the user promptly (within 30 minutes). The achievement of such an objective is aided by the above total system approach in which most of the data processing and transmission is automated. However, as we discuss later, the human forecaster is an important component of the nowcasting system and will remain so for the foreseeable future. By using an advanced man–computer interactive image display system it is possible to combine and manipulate the multiple data sources in such a way that the man can exercise his meteorological judgment rapidly and effectively within the context of the otherwise highly automated system (Browning, 1979).

3 Radar display format

We have come a long way since the days when forecasters had to rely on analogue PPI displays on which the plan positions of precipitation echoes were displayed qualitatively along with various kinds of unwanted echoes. We now have a computer at each site which controls the aerial scan, integrates the signal from the precipitation and converts it into the mean rainfall intensity within a rectangular matrix of squares of uniform size. The on-site computer also applies various corrections, including the removal of normal ground clutter, compensation for partial blockage of the beam by hills, and adjustments using telemetered raingauge data. The result is a conveniently digested stream of digital data which can be transmitted at regular intervals (5 or 15 min) along narrow bandwidth telephone lines to remote locations where the pattern of precipitation intensity observed by each radar can be portrayed as colours on a television monitor (Taylor and Browning, 1974) (see Fig. 4. (a), p. 52). For most practical purposes the display needs to be as simple as possible and we find it sufficient to display only the surface pattern of precipitation, and to use a coarse resolution of 5 km with 8 levels of intensity. Solid state technology has enabled the development of simple data stores which permit the rapid replay of successions of pictures. As a result the user can easily assess the motion and development of areas of rain. A system of this kind (Fig. 2) has been in use in the UK since the mid-70s. In addition to picture data (A in Fig. 2), rainfall totals are also available for use by hydrologist (B in Fig. 2). These totals are integrated at the radar sites for predefined time periods over multiple river subcatchments. Users may record all this information and replay picture sequences using an audio cassette recorder and modem (C in Fig. 2).

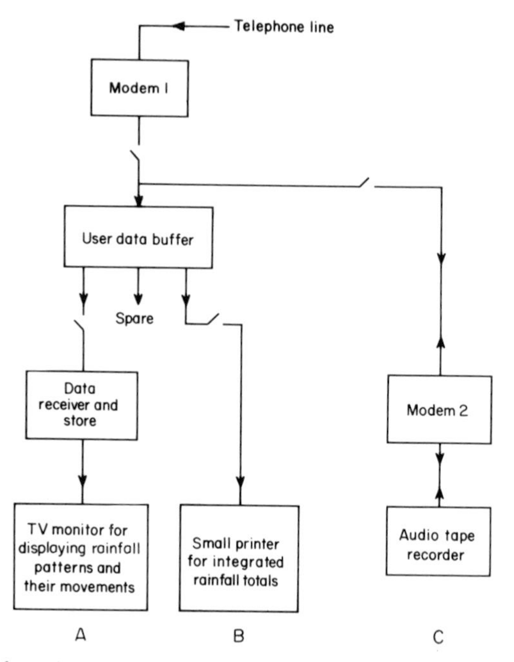

Fig. 2. Terminal equipment available to a user for receiving precipitation data by telephone line. The data may be received either direct from an individual radar or via the Mesoscale Analysis Centre, in which case the data will not only cover a larger area but can also consist of both actuals and forecasts. Examples of the types of pictures that can be displayed are given later in Fig. 4.

4 Radar network compositing

A single radar sees only a part – and sometimes a misleadingly small part – of a synoptic scale weather system so that it is often difficult to make sense of the fine detail it detects within the context of other traditional observations. For this reason, and because of the need to provide forecasts over larger areas, we employ a network of radars with overlapping coverage. Each radar sends its own preprocessed digital precipitation information by telephone to the Mesoscale Analysis Centre where a computer automatically derives a composite picture from all available radars within about 5 minutes of data time (Larke and Collier, 1981). At the time of writing (1981) the UK network consists of four radars. When all the radars function properly the composite picture is derived as in Fig. 3 (a). An example of a composite display is shown in Fig. 4 (b). (A single radar display is shown in Fig. 4 (a) for comparison.) If data from one or more radars are not received then the boundaries between radars are automatically redefined as shown in Fig. 3 (b)–(d). Clearly there is some useful redundancy

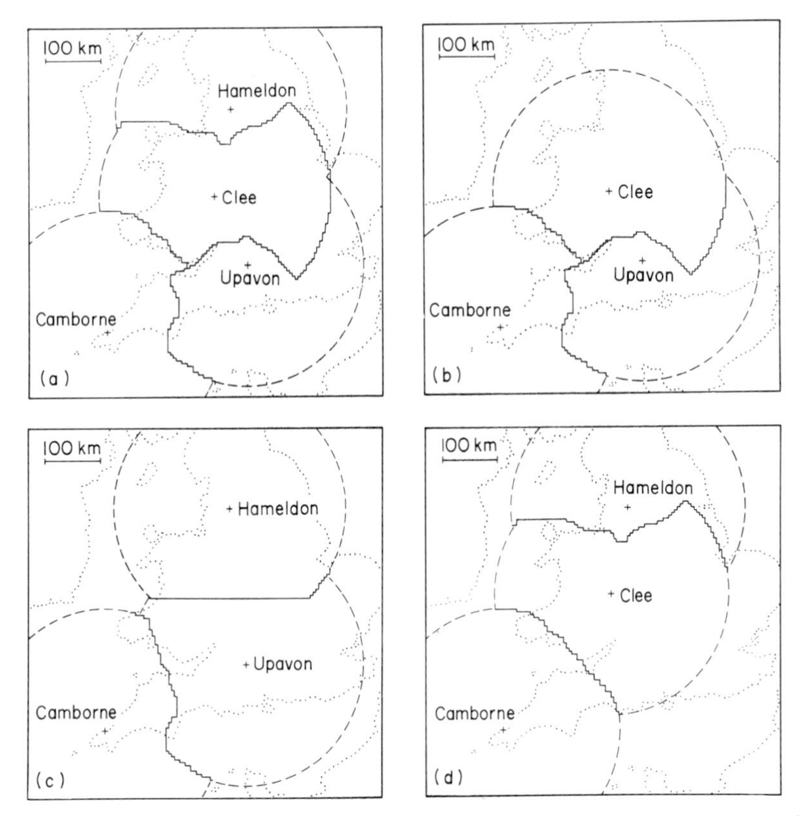

Fig. 3. Boundaries between radars used in forming the radar network composite pictures, shown for a selection of combinations of available radars. Dashed lines denote the maximum range (210 km) of individual radars the locations of which are indicated by crosses. Solid lines represent the designated boundaries.

and temporary loss of a radar is not disastrous. The boundaries between radars are predefined on the basis of siting considerations such that the data for any given area are obtained from the radar providing the best coverage; data are not smoothed across boundaries because any discontinuities that occur are regarded as helpful for identifying shortcomings in the data. Although the compositing is totally automated the forecaster at the Mesoscale Analysis Centre can intervene if necessary (see Section 7).

5 Satellite imagery

The geostationary satellite METEOSAT provides sufficiently frequent (30 min) cloud pictures to be useful for nowcasting purposes. Moreover, the spatial

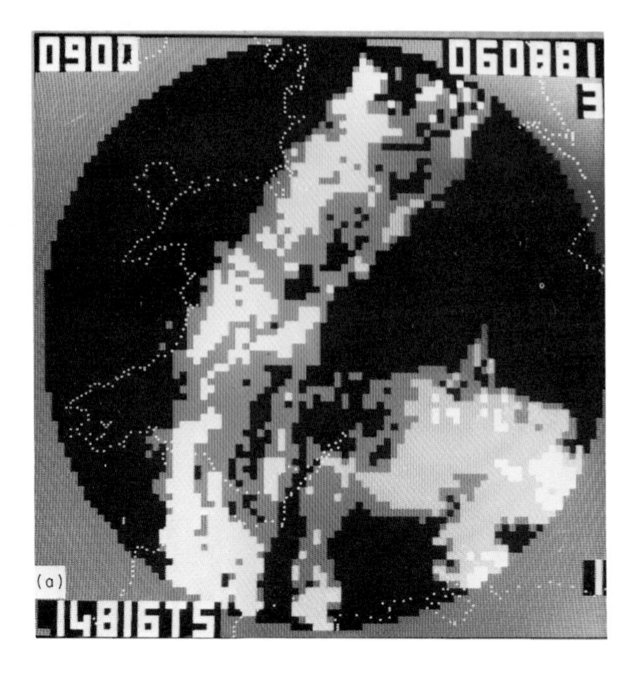
0900
060881
3
1
(a)
14816TS

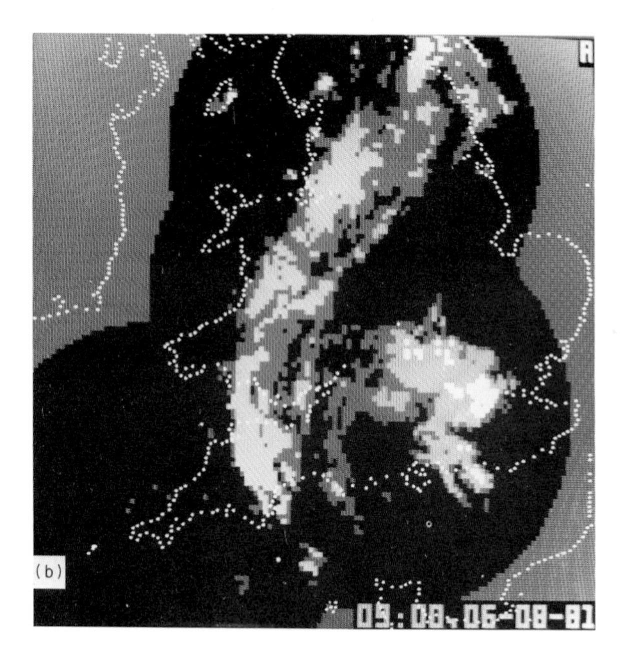
A
(b)
09:08,06-08-81

resolution over the UK is comparable with, or only marginally worse than, that (5 km) chosen to represent the radar data. The cloud images are available in the visible and infrared (also water vapour) channels. In simple terms the infrared radiance is a measure of cloud top height and the visible brightness a measure of cloud thickness. At night the infrared imagery can be used to give an indication of areas of precipitation but it requires subjective analysis using conceptual models of precipitation systems to identify and reject regions of non-precipitating high cloud. By day a computer-derived combination of visible and infrared imagery can be used. Although only a crude indication of precipitation, this gives a more reliable indication of probability of precipitation than either channel alone, especially when "calibrated" using radar data covering part of the same area (Lovejoy and Austin, 1979).

We use the METEOSAT imagery in two ways. The first is as an aid in the interpretation of the radar data, especially the analysis of precipitation patterns in areas of poor coverage. The second use of the imagery is to extend the coverage beyond the limits of the radar network so as to give advance warning of approaching precipitation systems. In order to make proper use of the satellite imagery it is automatically reprojected onto a Transverse Mercator format with 5 × 5 km cells identical to those of the radar network. Fig. 5 (a) shows an example of the reprojected data converted into probable precipitation intensity using the Lovejoy–Austin method. Fig. 5 (b) shows part of the same rain area just beginning to enter the coverage of the radar network.

6 Unusual error characteristics of the data

Our nowcasting system is aimed at deriving the surface rainfall intensity but, as pointed out above, the satellite does not measure this directly and it has to be

---

Fig. 4 (*opposite*). Precipitation distribution for 0900 GMT, 6 August 1981, displayed on a television monitor.

(a) Data from a single radar at Clee Hill, Shropshire (84 × 84, 5 km cells).

(b) Data from the UK network of four radars (128 × 128, 5 km cells).

The original displays were in colour and contained many intensity levels. For clarity the precipitation patterns in these monochrome pictures are shown as just three shades of grey, corresponding to light (< 1 mm h^{-1}), moderate to heavy (1–16 mm h^{-1}) and very heavy (> 16 mm h^{-1}) rainfall. Coastlines are shown electronically as white dots and the outer limit of radar coverage is represented by circles.

These pictures show organized regions of widespread rain with embedded mesoscale areas of heavy rain. Many thunderstorms were occurring, with a particularly intense one in the London area (the colour display showed that a rainfall rate exceeding 32 mm h^{-1} was occurring over an area of about 500 km^2 at this time). The storms were drifting northwards and a further intense storm can be seen over the Sussex coast which crossed London about 3 hours later.

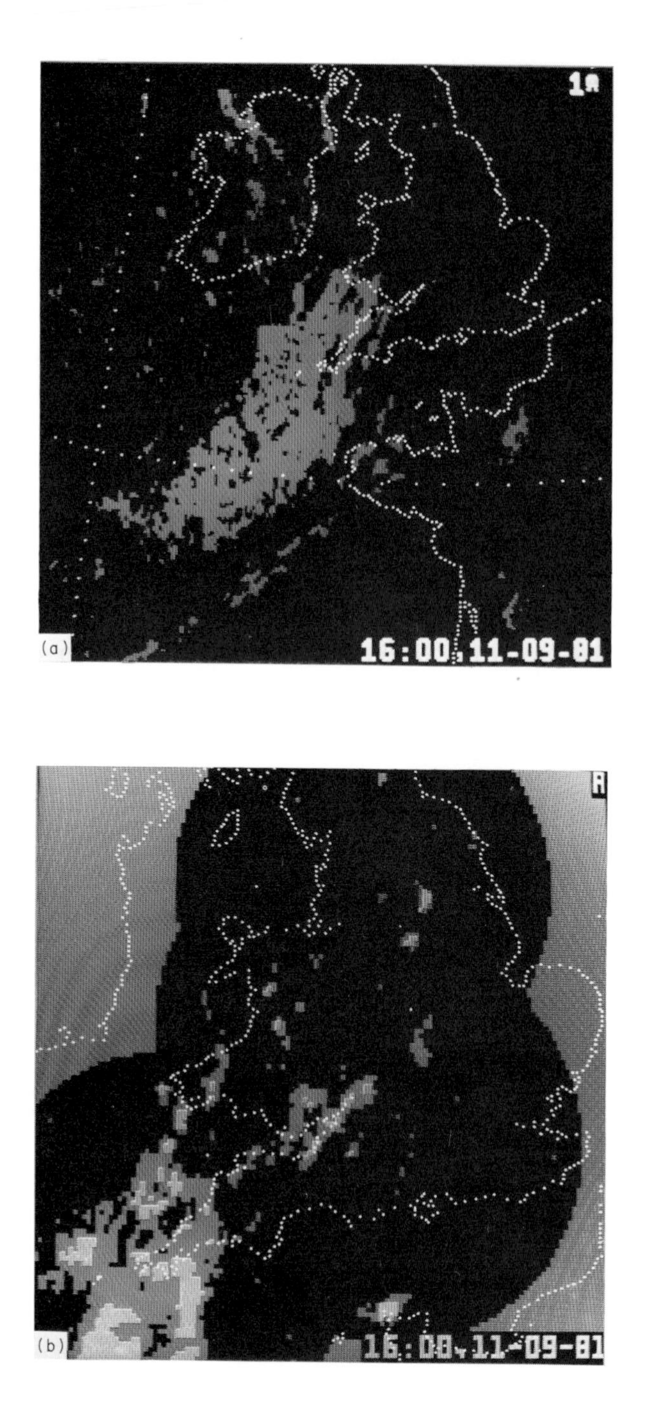

1A
(a)
16:00 11-09-81
A
(b)
16:00 11-09-81

inferred from the cloud imagery by a combination of objective and subjective analyses. The objective analyses may be regarded as having "errors" some of which can be identified using subjective meteorological judgment. There are similar difficulties also with the radar data because, although it detects precipitation, radar often fails to measure exactly what is wanted. Many experiments have been carried out to determine the accuracy of radar estimates of surface rainfall. One such experiment (CWPU, 1977) indicated an accuracy of 20% for hourly totals in subcatchments of area 100 km^2. However, accuracies as high as this can be achieved only under very restrictive conditions. In this project the accuracy was attained by limiting the range to 50 km, excluding cases when melting snow was in the beam, and using a high density of "calibration" raingauges (1 per 1000 km^2). In practical situations, especially where the radars are operated to long ranges (say 200 km), many problems arise which are not encountered in such carefully controlled experiments.

The radar measurement problems are illustrated in Fig. 6 which shows a cross-section through a commonly encountered type of mid-latitude frontal system, with a deep precipitation-bearing cloud layer overrunning dry air at the leading edge, followed by a region of increasingly shallow precipitation. Six sources of error in the measurement of surface rainfall are identified in Fig. 6. Most of the errors arise because the radar beam observes precipitation some distance above the ground; it is not solely a problem of variable drop size distribution. The feature common to all these errors is that they are due to variations in the physical nature of the phenomenon rather than to straightforward instrumental errors.

As far as possible these problems are dealt with automatically at the radar site preprocessing stage. For example, it is worthwhile using a few telemetering gauges to adjust the radar data on-site, provided the domains over which the

---

Fig. 5 (*opposite*). Precipitation distribution for 1600 GMT on 11 September 1981 displayed on a television monitor.

(a) Extent of the area with greater than 50% probability of surface rain, as derived objectively from a combination of infrared and visible imagery from METEOSAT 2 calibrated by radar using the method of Lovejoy and Austin. (256 × 256, 5 km cells: same projection as for radar network data but covering × 4 greater area).

(b) Rainfall patterns from the UK network of four radars. Dark and light shades of grey represent light-to-moderate and moderate-to-heavy rain, respectively. (128 × 128, 5 km cells.)

The original displays were in colour and contained many more intensity levels. Coastlines are shown by white dots.

The radar network picture in (b) shows an area of rain associated with a newly developing trough entering the zone of radar coverage; the satellite picture in (a) indicates the probable overall extent of this rain area.

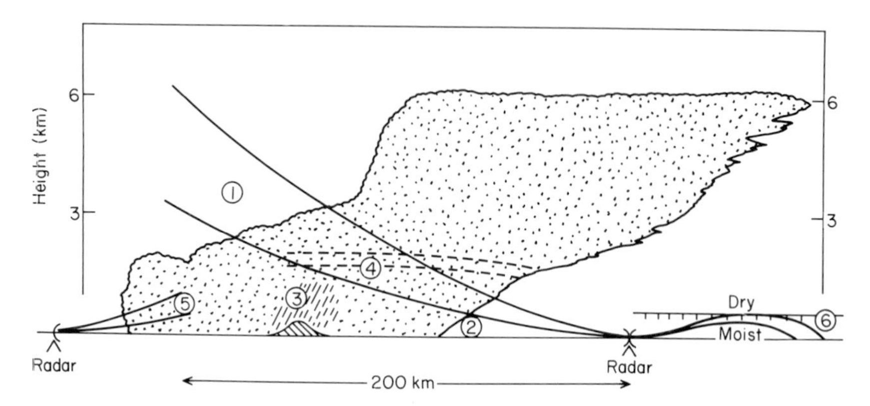

Fig. 6. Cross-section through an area of frontal precipitation illustrating six sources of error in the radar measurement of surface rainfall intensity, namely: (1) radar beam overshooting the shallow precipitation at long ranges; (2) low-level evaporation beneath the radar beam; (3) orographic enhancement above hills which goes undetected beneath the radar beam; (4) anomalously high radar signal from melting snow (the bright band); (5) underestimation of the intensity of drizzle because of the absence of large droplets; and (6) radar beam bent in the presence of a strong hydrolapse causing it to intercept land or sea.

adjustments apply are carefully selected. (This requires a physical approach, rather than a purely statistical one, to take into account orographic effects and synoptic type.) However, such corrections achieve only a limited improvement; it is the authors' opinion that the problems depicted in Fig. 6 are unlikely to be mastered by objective means alone. Instead a combination of objective and subjective methods is required based upon an analysis of the nature of the phenomenon in the light of other meteorological information. It is, however, unrealistic to expect outstation forecasters and other users to identify and correct these errors; they will be fully stretched in trying to absorb the full complexity of the actual weather situation on the mesoscale without having to be experts in radar and satellite meteorology as well. Rather, it is better for these tasks to be carried out once and for all at the Mesoscale Analysis Centre. Here a specialist can carry out the necessary analyses by manipulating the radar, satellite and other data on a versatile interactive display system.

7 The interactive display system and the central role of the forecaster

The meteorological use of man–computer interactive video display systems was pioneered at the University of Wisconsin (Chatters and Suomi, 1975). The interactive system that we have devised, known as the FRONTIERS system, has been tailored to the specific needs of analysing and forecasting precipitation

(Browning, 1979; Saker, 1981). So that it can be operated rapidly in a forecasting environment, close attention has been paid to its ergonomic design. The controls are simple and direct. The response is virtually instantaneous. Above all, the technicalities of the manipulation of the large data sets are transparent to the operator, so that he is free to concentrate on the meteorology. The different data sources can be accessed, animated, superimposed, and also modified, by simply pointing a finger at the image display and at a menu on a neighbouring display.

We now list some of the ways in which the central interactive display will be used. The list is illustrative rather than comprehensive; nevertheless, it might seem from what follows that the overall analysis would be an unduly time-consuming task. In fact it is not as onerous as it might seem because on any given occasion it is unlikely that all the problems would require detailed attention. Moreover the nature of the analysis will tend not to change rapidly from one analysis cycle to the next, and so there is scope for applying corrections which are "the same as before".

7.1 SOME STEPS IN THE RADAR ANALYSIS

i. Spurious (non-precipitation) echoes can be automatically rejected at the radar sites using various techniques but in the very nature of things some spurious signals still get through in practice. It is clearly important to remove these from the data that are sent to the users because of the damage they can do to the credibility of the product. There are two main causes of spurious echoes: interference, and ground or sea echoes, especially during anomalous propagation conditions. They may be identified subjectively in several ways: from their non-meteorological appearance, or (with the aid of animation) from any deviation in their motion from that of the steering level winds, or (with the aid of geographical overlays) from their location in favoured regions. Alternatively the spurious echoes may show up on one radar but not on another radar covering the same region, or the satellite may show that they occur in a region of no cloud. However, none of this evidence can be relied on by itself and human judgment is needed to weigh each strand of evidence. Echoes diagnosed as spurious can then be eliminated simply by drawing a line around them on the display.

ii. Although data from each radar are adjusted on-site using telemetered rain-gauge information, there are several reasons why individual "calibration" gauges may be unrepresentative. Such gauges may be identified on the interactive display by superimposing the gauge values directly on the radar pattern together with other diagnostic information. New radar adjustments can then be derived automatically with these gauges excluded.

iii. Shallow precipitation may totally escape detection at long ranges and this leads to gaps in coverage between radars. Such gaps may be filled in by the analyst by interpolating between individual radars in the light of guidance from the satellite imagery or by extrapolating the position of radar rainfall echoes previously observed upwind.

iv. Orographic enhancement of frontal rain occurs mainly in the lowest kilometre of the atmosphere and so usually it is underestimated by radar. A useful indication of the orographic enhancement can be obtained by using mesoscale climatological fields of rainfall enhancement given as a function of parameters such as low-level wind velocity and relative humidity (Browning and Hill, 1981). Although the orographic corrections can be applied automatically to the radar rainfall patterns, the selection of the parameters on an area-by-area basis is best done subjectively in the light of conventional synoptic data displayed as an overlay to the radar pattern.

7.2 SATELLITE ANALYSES

i. The satellite imagery is normally preregistered on receipt. However, errors of ten or more kilometres may remain which, although insignificant for synoptic analysis, are important for local forecasting. Precision registration is achieved using the interactive display by lining up visible coastlines manually against a suitable overlay. The imagery can be enhanced to reveal the land–sea boundaries by manually adjusting the contrast and colour allocation.

ii. Rainfall probability maps derived automatically from satellite imagery will tend to exaggerate the extent of surface precipitation where there are large areas of high frontal cloud, and underestimate it in some regions of shallow convection. The analyst can identify some of these errors on the basis of other strands of observational evidence together with conceptual models of frontal morphology. It is then a simple matter for him to modify the rainfall analysis directly on the screen itself. Needless to say the analysis becomes increasingly unreliable the further one strays from the central area of radar network coverage.

7.3 FORECASTING BY LINEAR EXTRAPOLATION OF THE RADAR–SATELLITE DATA

The nowcasting system can be run in a totally automated mode, bypassing the subjective analyses described above and using a purely objective extrapolation technique to derive very-short-range forecasts. However, an important factor contributing to the disappointing results sometimes obtained using objective extrapolation techniques is thought to be the limitations in the initial data. The steps described above will go a long way towards overcoming these difficulties.

Shortcomings in the objective extrapolation techniques themselves can also be addressed using the interactive display in one of the following ways:

i. An objectively derived forecast (or the parameters of the forecast procedure) can be modified in the light of a subjective assessment of the movement of the rainfall patterns obtained by replaying a sequence of pictures.
ii. The velocity of parts of the rainfall pattern can be determined entirely by manual means (for example by using a Lagrangian replay technique in which the pattern velocity is adjusted to keep recognizable features stationary within selected windows).

The essence of the nowcasting approach is the linear extrapolation of the current situation. In regions where sudden outbreaks of deep convection are common such procedures will obviously be unsuitable. However, in regions such as the UK where the weather is dominated by frontal disturbances simple advection is a valuable forecasting tool at least for a few hours ahead. Development and decay can to some extent be taken into account subjectively by the forecaster using conceptual models and other observations.

8 Future developments

The next stage in improving short-range forecasts of precipitation will be to develop ways of assimilating the nowcast information into a mesoscale numerical prediction model. However, information of the kind discussed in this chapter is only one source of nowcast data for initializing a numerical model. Accordingly we would hope to see the development of multiple interactive work stations at the Mesoscale Analysis Centre, perhaps as shown in Fig. 7. Work Station 1, in addition to providing precipitation nowcasts, would enable fields of humidity and divergence to be inferred. Work Station 2 would take inputs from a satellite sounder and conventional sources so as to provide nowcast fields of temperature and humidity. It would also take inputs from satellites and conventional sources to provide nowcast wind fields. All of these nowcast fields would then need to be reconciled with one another and with forecast background fields before they can be used in the mesoscale prediction model; this would be achieved in Work Station 3. Much research will be required to develop these procedures. Displaced real-time forecast evaluation procedures of the kind described by Beran and MacDonald (1982) will be required to optimize the total system design. In general, however, as the period of the forecast increases beyond 6 hours, the value of having good initial data on the mesoscale will tend to decrease and the relative importance of synoptic scale and topographic forcing will increase.

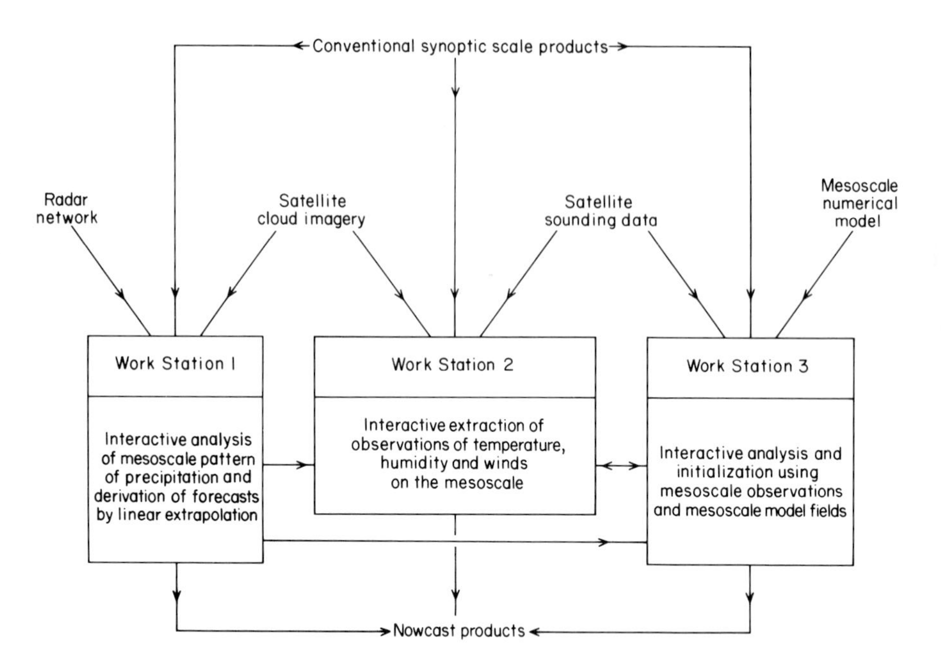

Fig. 7. Possible configuration of video work stations in a future nowcasting and very-short-range forecasting centre.

References

Beran, D. W. and MacDonald, A. E. (1982). This volume, pp. 17–23.

Browning, K. A. (1979). The FRONTIERS plan: a strategy for using radar and satellite imagery for very-short-range precipitation forecasting. *Met. Mag.* **108**, 161–184.

Browning, K. A. (1980). Radar as part of an integrated system for measuring and forecasting rain in the UK: progress and plans. *Weather* **35**, 94–104.

Browning, K. A. and Hill, F. F. (1981). Orographic rain. *Weather* **36**, 326–329.

Chatters, G. C. and Suomi, V. E. (1975). The applications of McIDAS. New York IEEE Trans. Geosci. Electr., GE13, pp. 137–146.

Collier, C. G. (1980). Data processing in the Meteorological Office Short-period Weather Forecasting Pilot Project. *Met. Mag.* **109**, 161–177.

CWPU (1977). Dee Weather Radar and Real Time Hydrological Forecasting Project, Report by the Steering Committee, Central Water Planning Unit, London, HMSO.

Larke, P. R. and Collier, C. G. (1981). Merging data from several radars. Preprint Vol. COST 72 Workshop/Seminar on Weather Radar, ECMWF Reading and Meteorological Office, Radar Research Laboratory, Malvern, 9–11 March 1981, pp. 108–123.

Lovejoy, S. and Austin, G. L. (1979). The delineation of rain areas from visible and IR satellite data for GATE and mid-latitudes. *Atmosphere-Ocean* **17**, 77–92.

Saker, N. J. (1981). The design of the FRONTIERS Interactive Display System. *In* "Nowcasting: mesoscale observations and short-range prediction", pp. 357–361. Proceedings of a symposium at the IAMAP General Assembly, 25–28 August 1981, Hamburg. European Space Agency, ESA SP-165.

Taylor, B. C. and Browning, K. A. (1974). Towards an automated weather radar Network, *Weather* **29**, 202–216.

PART 2

New Forms of Observations

Introduction

The new opportunities for improved very-short-range forecasts have been stimulated to a large extent by advances in observational capabilities, especially remote probing techniques. Some of these capabilities, such as satellite cloud imagery and conventional ground-based radar, are well established in themselves, and the challenge is how best to use the data for forecasting. These aspects are considered elsewhere in this volume. There are, however, many other mesoscale observational techniques which may become available for operational use over the coming decade and we deal with some of these in Part 2.

Little (Chapter 2.1) presents a wide range of ground-based remote probing techniques likely to find applications in nowcasting. These include laser, sodar and radar techniques for measurements in the boundary layer. Even more important is the possibility of using VHF/UHF Doppler radar and microwave radiometer techniques, continuously and unmanned, to provide wind, temperature and humidity profiles over a considerable depth. Though lacking the vertical resolution of radiosondes, this ground-based sounding system, when used in conjunction with satellite-borne soundings, will provide good horizontal and temporal coverage, and shows promise of being able to fulfil many of the requirements for forecasting.

One particular remote probing technique that has been used in research for many years, and is now being developed for operational use, is microwave Doppler radar. This is discussed by Wilson and Wilk (Chapter 2.2). Although such systems may not be cost effective in the short run in those parts of the world where severe convective storms are uncommon, there can be little doubt that Doppler radar provides the most effective means for detecting such major wind hazards as tornadoes, downbursts and gustfronts.

Two chapters (2.3 and 2.4) deal with the capability of satellite-borne sounders to observe the 3-D temperature and humidity structure of the atmosphere with mesoscale spatial resolution. The first of these chapters, by Kelly and his co-workers in Australia, deals with the data from the TIROS Operational Vertical Sounder (TOVS) provided by the TIROS-N type of polar-orbiting satellites. There is obviously strong motivation for this kind of data to be exploited in the Southern Hemisphere where other forms of sounding data are scarce; however, it is evident that direct-readout TOVS data provide a degree of detail horizontally on the mesoscale that would be impossible to match by other (non-satellite) means, even in the Northern Hemisphere.

Unfortunately the data from just one or two polar orbiting satellites are available too infrequently for most local forecasting purposes. To overcome this limitation the so-called VAS system has been developed in the United States for use on a geostationary satellite to provide soundings with mesoscale spatial resolution (75 km) at 1-hour intervals. Smith and his collaborates (Chapter 2.4) demonstrate the capability of such a system to observe the pre-storm environment prior to the development of severe convective storms, at least when large areas of clear sky prevail before the outbreak of severe weather. The results presented are of a preliminary nature and a number of problem areas remain to be investigated. Nevertheless early indications are that this approach has considerable potential for the prediction of severe local storms.

We round off Part 2 with a contribution from Shapiro and co-workers showing how ozone measurements from the Nimbus 7 satellite can reveal features of upper tropospheric jet streams with a degree of horizontal resolution that promises to be of practical value, for example in the planning of more economical aircraft flight tracks.

This book is illustrative rather than comprehensive and there is no discussion of the present use of satellite cloud displacements for inferring wind fields. Neither do we consider exciting but more futuristic possibilities such as the use of satellite-borne lasers for measuring 3-D wind fields in clear air, or satellite-borne microwave sounders for measuring surface pressure.

K. A. Browning

2.1

Ground-based Remote Sensing for Meteorological Nowcasting

C. GORDON LITTLE

1. Introduction

In preparing very-short-range local weather forecasts, we are attempting to predict the near-term evolution of a three-dimensional fluid affected by processes and phenomena active on many time and space scales. To be able to predict the future state of the atmosphere we must be able to monitor its current state. However, an adequate definition of its current state requires frequent measurements in all three spatial dimensions. While acquisition of two-dimensional data sets across the surface of the ground is often possible by means of arrays of *in situ* instruments, such instruments do not provide a practicable method of obtaining continuous data at heights of more than a few tens of metres. To meet this need for atmospheric measurements aloft, meteorologists are increasingly turning to remote-sensing methods, in which middle and upper atmospheric parameters are measured remotely, using ground-based sensors. Such methods potentially have many advantages over equivalent arrays of *in situ* sensors, including the fact that conceptually the data are available

- in one, two, or three spatial dimensions,
- without the use of towers, balloons, or aircraft,
- without the use of expensive telemetry networks,
- with excellent continuity in space and time,
- with excellent resolution in space and time,
- as spatial averages,
- and using systems that may be readily automated.

This chapter summarizes progress in creating remote sensors capable of providing at least some of these advantages to those concerned with operational nowcasting and forecasting.

2. Surface (or near-surface) measurements

During the past decade, research on the interaction of visible and infrared waves with the atmosphere has resulted in development of several remote sensors based on the scintillation of optical signals. Two such sensors, the transverse wind sensor (Clifford *et al.*, 1975; Ochs *et al.*, 1976; Ochs *et al.*, 1979; Wang *et al.*, 1981), and the laser weather identifier (Earnshaw *et al.*, 1978; Wang *et al.*, 1979), have applications to nowcasting.

2.1 TRANSVERSE WIND SENSOR

The phenomenon of scintillation, or twinkling, of stars, has been observed for centuries. It is now known to be caused by the small-scale (of the order of ~ 1 cm) fluctuations in optical refractive index of the air due to small changes in temperature, and hence, of density. As the optical wave propagates through this slightly inhomogeneous, fluctuating atmosphere, refraction and diffraction produce a non-uniform, fluctuating distribution of light across any receiving plane. If there is a transverse component of the wind across the optical path, this diffraction pattern is observed to move across the receiving plane with a direction and speed proportional to the transverse component of the wind. Hence, the measurement of the speed of motion of the diffraction pattern permits measurement of the mean value of transverse wind speed, averaged along the length of the optical path.

Early versions of such a transverse wind sensor made use of a small continuous wave (CW) laser. More recent theoretical and experimental work has shown that larger aperture receivers and transmitters, using light that is less coherent than laser light, are preferable, in order to avoid erroneous wind speed measurements during very turbulent conditions when the amplitude of the scintillations may saturate (Ochs *et al.*, 1976). Various methods of computing the path-averaged transverse velocity from the scintillations observed at two horizontally spaced receiving apertures are discussed by Wang *et al.* (1981); these authors show that the most satisfactory method makes use of the full time-lagged cross-covariance curve, rather than the time delay at peak correlation, or the slope at zero-time delay.

Figure 1 compares optical scintillation measurements of transverse wind speed with the average of 6 anemometers spaced uniformly along the 300-m optical path. Although early systems required that electrical power be available at each end of the optical path, recent systems (Ochs *et al.*, 1979) make use of modulated light-emitting diodes at 0.94 μm powered by a small lead–acid battery charged by solar cells. The optical signal can be used as a telemetry channel, as well as a method of measuring the path-averaged transverse wind speed.

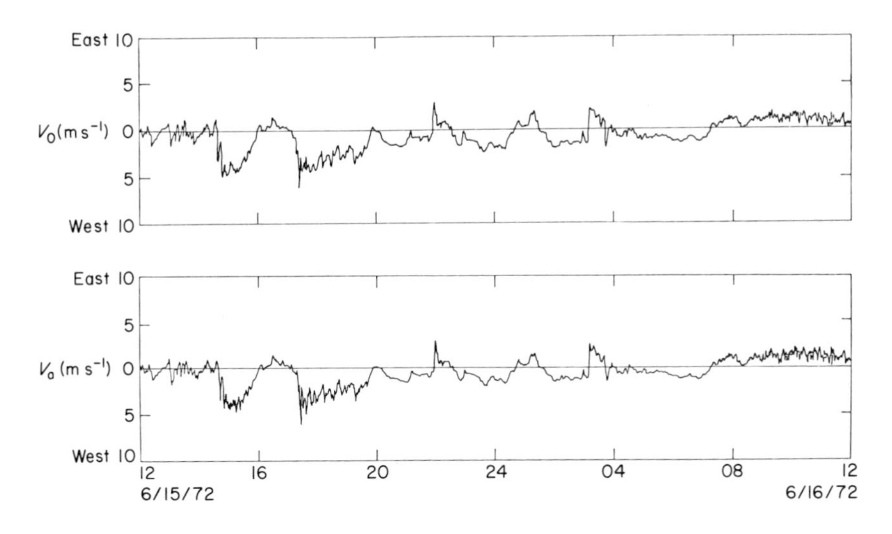

Fig. 1. Comparison of average transverse wind speed measured optically (V_0) and by 6 anemometers spaced uniformly along a 300-metre optical path (V_a).

These path-averaging transverse wind sensors have found use in the measurement of area-averaged wind speed and direction (using two approximately orthogonal paths) and convergence (using three paths forming a closed triangle). Single-ended versions, using scintillations of the natural scene, have also been developed (Clifford *et al*., 1975). The transverse wind sensors are typically used on path lengths of roughly 1 to 10 km. They have been found particularly valuable in measuring very low wind speeds (unlike mechanical systems, they have no minimum start-up speed), and in measuring path-averaged drainage wind speeds in mountain valleys, at various heights above the valley floor.

These instruments have been found to be highly reliable, and, once installed, can operate for many months without attention. The one significant limitation is that measurements are not available under conditions of fog or significant precipitation because of heavy attenuation of the optical beam.

2.2 LASER WEATHER IDENTIFIER

The second instrument based on the use of optical scintillations is the laser weather identifier. This device operates automatically over short paths at remote locations to identify the local surface weather conditions, especially the occurrence and nature of precipitation.

The scintillation of optical signals during precipitation is quite different from that observed under clear conditions. Mr. Lawrence and his colleagues (notably K. B. Earnshaw and Drs Ting-i Wang and S. F. Clifford) have explained these differences and have shown how measurements of the vertical velocity of the diffraction pattern can be used to measure path-averaged rainfall rate and path-averaged drop-size distributions (Earnshaw *et al.*, 1978; Wang *et al.*, 1979).

Their work has shown that the scintillations produced by advection of small temperature eddies across the optical path are of much lower frequency than those produced by rain or hail. They have demonstrated that it is possible to use this difference in scintillation frequency as a simple indicator of the presence of rain or hail. Unfortunately, this simple approach cannot be used in every case to identify snowfall, because of an overlap between scintillation rates observed during snow and those observed in clear air during high winds. This ambiguity can be resolved by measuring the vertical velocity of the diffraction pattern, which averages to zero for temperature eddies, but has a mean downward velocity during snowfall. The laser weather identifier they have built consists of a He–Ne laser whose beam is expanded by a 10-cm diameter telescope and directed over a 60-m path to the receiving system. This consists of two detectors, each receiving energy from a horizontal 10 cm × 1 mm observing slit, the slits being spaced vertically 1.5 cm apart. The time delay of the scintillation output of the lower detector relative to the upper detector is used to determine the fall velocity of any precipitation along the path, and hence to differentiate between rain, snow, and large hail. In addition, the scintillation power at frequencies centred on 1 kHz is found to be a good measure of path-averaged rainfall rate. For a well-calibrated system, the accuracy is believed to be within 10% over the range 1.0 to 100 mm h^{-1}.

3 Boundary layer measurements

3.1 ACOUSTIC ECHO SOUNDING

Acoustic waves interact strongly with the atmosphere, and relatively simple equipment can therefore be used to measure the interactions, and hence the related properties of the atmosphere. As a result, the acoustic echo sounder (or sodar) is the most commonly used remote sensor for boundary layer studies. (See Brown and Hall, 1978, for a detailed review of this field.)

Acoustic waves propagating through the atmosphere are scattered most strongly by small-scale turbulent fluctuations in temperature and air speed (scatter from humidity fluctuations is usually not important over land). In the monostatic (backscatter) mode, i.e. using co-located transmitters and receivers, the sounder is sensitive to temperature eddies (assumed isotropic) of size equal to half the acoustic wavelength (typically, the acoustic wave lengths used are between 20 and 50 cm). For scattering angles other than $\theta = 180°$, the scattering occurs from both temperature and velocity fluctuations of spatial size $\lambda/2$ sin

($\theta/2$). Thus a monostatic system can be used to prepare height-time plots of the intensity of the temperature eddies; a bistatic system adds information on the height-time variation of atmospheric turbulence. In addition, any motion of the scatterers that changes the round-trip path length from the transmitter to the scatterer and back to the receiver results in a Doppler shift of the observed acoustic frequency that can be used to derive information on the velocity of the atmospheric eddies responsible for the scattering. Since these eddies travel with the air velocity, acoustic echo sounders can be used to measure profiles of mean wind speed and direction. The intensity of the turbulence as a function of height can be determined by observing the fluctuations in Doppler shift of echoes from a fixed height as a function of time, or the differences in Doppler shift of echoes observed simultaneously at different ranges. Tests under the auspices of the World Meteorological Organization (Kaimal *et al.*, 1980) have shown that 20-minute averages of the u or v components typically differ by less than 1 m s^{-1} relative to *in situ* sensors at the same height on a neighbouring meteorological tower (see Fig. 2). Some of this difference is attributable to the 300-m separation of the sounder from the tower.

Experience has shown that the acoustic echo sounder can be a good indicator of the thickness of the atmospheric boundary layer (often termed "the mixing depth"). (See Fig. 3.) Since mixing depth and horizontal velocity are required to define the transport and diffusion of pollutants, acoustic echo sounders have found a major role in air pollution studies. They are frequently used in the site selection and plant design of industrial sources of pollution such as power plants or smelters; once the plant is installed, the acoustic sounder data may be used in real time to facilitate plant operations, or to identify harzardous regions in the event of an accidental release of toxic materials.

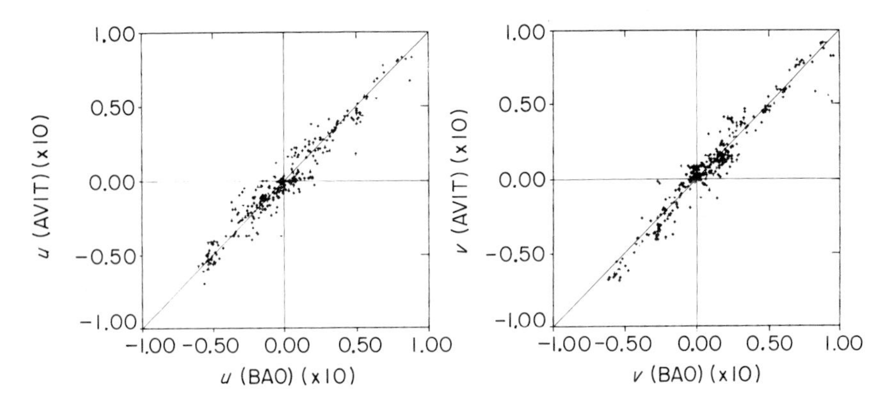

Fig. 2. Comparison of acoustic Doppler (ordinates) and meteorological tower (abcissas) measurements of u and v components of horizontal wind. The velocity scale in each diagram is ± 10 m s^{-1}.

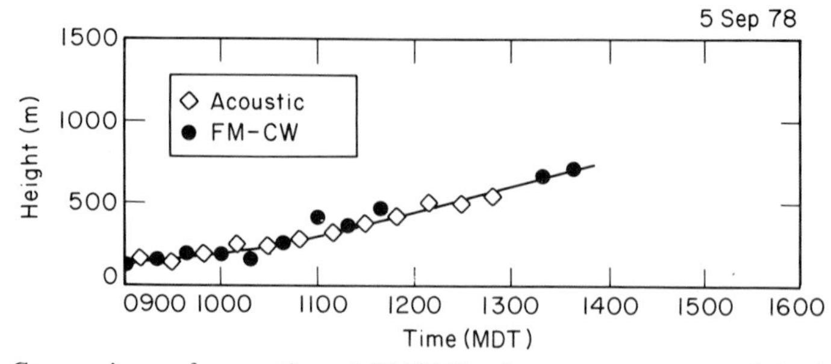

Fig. 3. Comparison of acoustic and FM/CW radar measurements of the height of the inversion layer, as a function of time. The two sounders were separated by about 700 metres.

Although useful for boundary layer studies (especially those related to air pollution) acoustic echo sounders have serious limitations. Their maximum height is usually limited to about 500 m; their performance is seriously degraded by increased noise levels during periods of high wind or rain – though these, of course, are rarely a problem in air pollution conditions. Also, the slow propagation velocity of acoustic waves precludes their use to scan the surrounding region in a radar-like mode to give volume coverage.

3.2 ELECTROMAGNETIC ECHO SOUNDING

Electromagnetic echo sounders can be used over a wide range of wavelengths to probe the atmosphere. At near-ultraviolet and visible wavelengths, the Rayleigh scatter from atmospheric molecules is sufficiently strong to be detected. Typically, however, at these and at infrared wavelengths, any clear-sky boundary layer echoes are likely to be from aerosol particles, rather than from atmospheric molecules. At radio wavelengths, molecular Rayleigh scatter and aerosol scatter are not detected, though insects are often observed at the shorter radar wavelengths. At longer wavelengths ($\gtrsim$ 10 cm) clear air echoes are usually due to small-scale ($\lambda/2$) fluctuations in radio frequency atmospheric refractive index, produced by inhomogeneities in atmospheric temperature and humidity. Cloud and precipitation particles scatter electromagnetic waves quite strongly, especially for wavelengths comparable with or smaller than the hydrometeors.

In addition, and particularly at infrared and shorter wavelengths, the interaction of electromagnetic waves with atmospheric molecules can become important, leading (for example) to absorption and scatter (e.g. Raman scatter) that is highly dependent upon chemical constituent and electromagnetic frequency.

Despite the very rich and varied range of interactions of electromagnetic waves with the atmospheric boundary layer, relatively little operational use is made of active (radar-like) electromagnetic remote sensors of the clear-sky boundary layer. However, two Doppler systems appear to offer promise of operational use in the clear boundary layer: microwave Doppler radars, which obtain echoes from the $\lambda/2$ size refractivity fluctuations, and infrared (10-μm) Doppler lidars, which obtain echoes from the larger aerosol particles. Doviak and Berger (1980) have shown examples of boundary layer velocity fields measured by two large 10-cm wavelength Doppler radars over areas of some 600 km^2. Similar measurements with a single Doppler radar have been used by Koscielny *et al.* (1981) to map mesoscale divergence patterns within the cloud-free convective boundary layer (see also Wilson and Wilk, this volume, pp. 87–105). Such Doppler radars, or frequency modulated, continuous wave (FM/CW) Doppler radars of the type discussed by Strauch (1979), would be useful in the vicinity of airports to identify gust fronts and wind shear events, and possibly also wake vortices. Such clear-air radars have the advantage of essentially all-weather capability.

In clear air, recent advances in pulsed Doppler lidar (Lawrence *et al.*, 1980) offer promise of single-station three-dimensional mapping of aerosol and atmospheric velocity fields to ranges of 20 km, with extremely high angular resolution (better than 10^{-4} radian). Cloud heights and sky coverage by clouds would also be available from such a device.

Although various lidar spectroscopic methods have been proposed (and in some cases demonstrated) for the measurement of atmospheric temperature and humidity profiles, so far as the author is aware, none is in routine operational use for nowcasting purposes. Typically, such sensors fail in rain or fog; in other cases sensitivity, height range, and accuracy are not yet adequate, or costs (including manpower) and reliability to do not yet permit routine operational use. Lidar systems are, however, commercially available for monitoring cloud-base heights.

4 Vertical profiles of wind, temperature and humidity

Although near-surface and atmospheric boundary layer data are important to nowcasting, attempts to forecast more than 2 h in advance also require information on the state of the upper atmosphere surrounding the site. The area for which these upper atmosphere data are required is a function of the length of the forecast; a 12-h forecast will require upper air data over a domain of the order of 1000 km square. The standard, worldwide method of obtaining upper air data is the radiosonde station, which releases (typically once per 6 or 12 h) a small balloon carrying instrumentation that permits measurement of the profile of wind, temperature, and humidity. Thus, the basic upper air data over the United States used daily by the US National Weather Service is obtained once per 12 h at approximately 70 sites across the contiguous 48 states, i.e. at horizontal spacings of roughly 350 km.

Similar networks of radiosonde stations are basic to the weather services of the world. Nevertheless, the radiosonde does have some disadvantages:

- profiles are not available continuously in time (typically, one sample per 6 or 12 h);
- manpower requirements are significant;
- accuracies (particularly of wind and humidity profiles) are less than desired.

Therefore, on the basis of about a decade of research on active and passive atmospheric remote sensors, the NOAA Wave Propagation Laboratory some two years ago began the development of an automatic unattended, ground-based, continuous profiler of winds, temperature, and humidity. This work has been conducted under the direction of Dr D. C. Hogg.

Certain desirable features for the PROFILER were identified at the outset of the program. Specifically, so far as practicable, the data should be available continuously in time; the system should operate automatically, i.e. without an attendant (except for occasional maintenance and servicing); and the profiles should always be available at least to tropopause heights, with accuracies and height resolutions consistent with the needs of numerical weather prediction models. Finally, the system should have excellent, long-term reliability.

The basic components of the prototype system are illustrated in Fig. 4. The function and nature of each main building block are discussed in turn below.

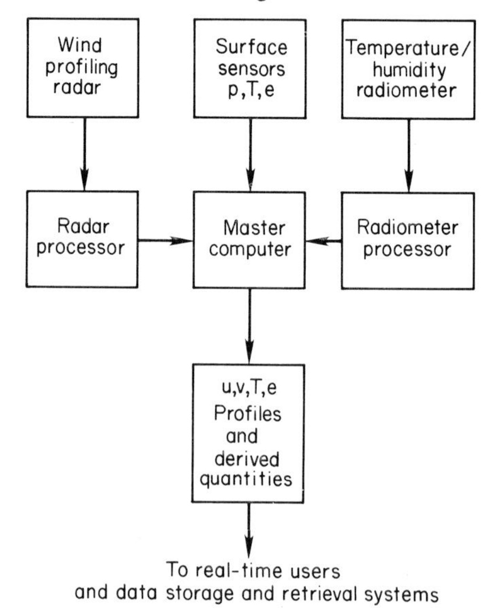

Fig. 4. Diagram illustrating the major components of the WPL PROFILER.

4.1 WIND PROFILING SYSTEM

Two versions of clear air radar are being evaluated for use in the PROFILER. The first is based on the early work of Woodman and Guillen (1974), who showed that the large 50-MHz ionospheric Thomson scatter radar at Jicamarca, Peru, could be used to obtain echoes and measure winds at upper tropospheric and lower stratospheric heights. This work has been followed up vigorously at other locations by the NOAA Aeronomy Laboratory (see Balsley, 1981, for a review of this rapidly developing field).

The PROFILER currently uses the Platteville, Colorado, 50-MHz radar, developed originally by the Aeronomy Laboratory as a prototype module for the large Poker Flat, Alaska, MST (Mesosphere, Stratosphere, Troposphere) radar. The radar (see Fig. 5) has operated almost continuously since October 1980. Recently a third, fixed, zenith-pointing antenna (not shown in Fig. 5) has been added to aid in identification of tropopause height, and in the measurement of vertical velocities. Current parameters of the radar are shown in Table 1; it is planned in the next year or so to improve the minimum height and height resolution to 0.5 km, and to increase the radiated power by about 10 dB.

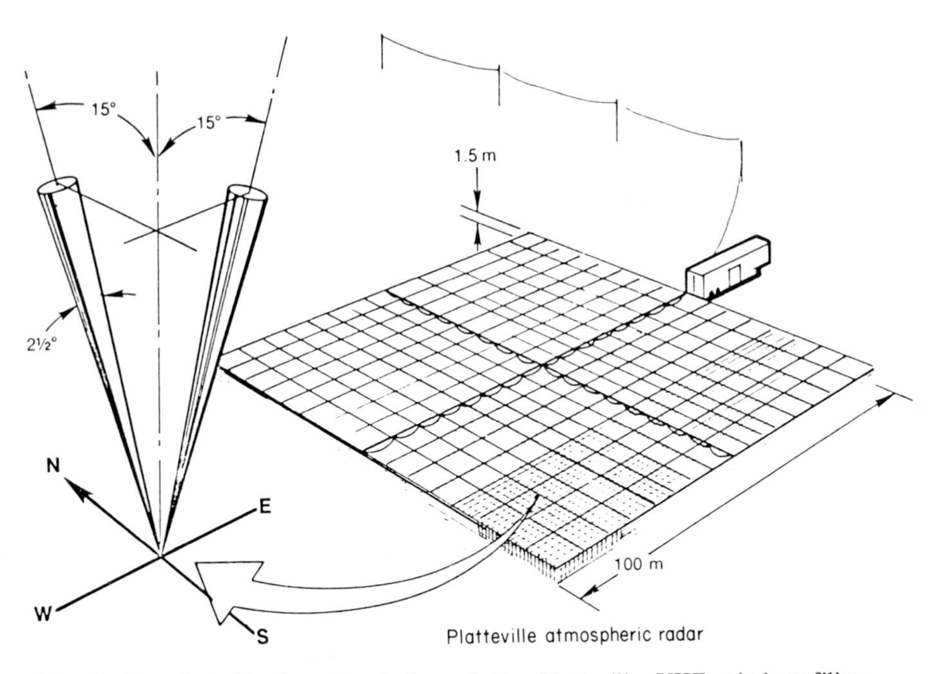

Fig. 5. Sketch indicating the design of the Platteville VHF wind-profiling radar.

TABLE 1

Radar parameter	VHF radar	UHF radar
Frequency	49.92 MHz ($\lambda_0 = 6.0054$ m)	915 MHz ($\lambda_0 = 33$ cm)
Pulse width	16 μs	Variable, 0.6–4.8 μs
Pulse period	2400 μs	Variable, 20–500 μs
Antenna size	100 m × 100 m (3 arrays)	10 m × 10 m (3 feeds)
Beam positions	Zenith; 15° off zenith to north; 15° off zenith to east	Zenith; 15° off zenith to north; 15° off zenith to east
Peak power	12 kw (each beam)	6 kw
Average power	80 W (each beam)	1500 W maximum
Minimum receiving range	2.4 km AGL	~ 100 m
Receiver range spacing	1500 m	Variable, ⩾ 100 m
Number of range locations	13 (2.4–20.4 km AGL)	32
Mode 1 (for horizontal winds)		
Maximum horizontal wind speed	± 75.46 ms^{-1}	~ ± 70 ms^{-1}
Horizontal wind resolution	± 0.29 ms^{-1}	< ± 0.25 ms^{-1}
Dwell time	130 s	Variable, 10–90 s
Mode 2 (for vertical winds)		
Maximum vertical wind speed	± 2.44 ms^{-1}	~ ± 10 ms^{-1}
Vertical wind resolution	± 0.01 ms^{-1}	~ ± 0.04 ms^{-1}
Dwell time	90 s	Variable, 10–90 s

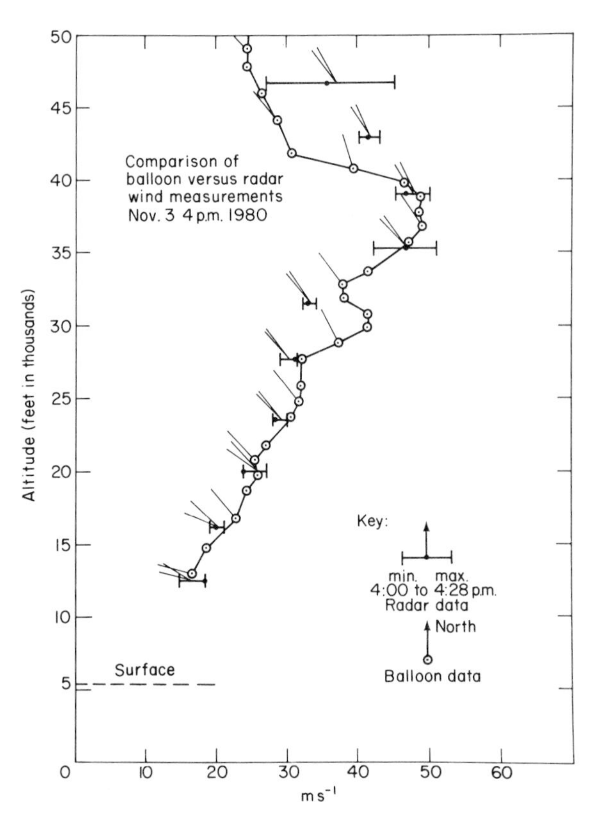

Fig. 6. Comparison of radiosonde and VHF wind profiles obtained simultaneously some 80 km apart.

A comparison of the Platteville measurements of winds with those obtained approximately simultaneously by the Denver radiosonde is given in Fig. 6. Note that the figure includes the full range (*not* the r.m.s. deviation) of the seven values of wind speed and direction obtained at each height interval during the 28 minutes covering the radiosonde flight. In general, the agreement is good, especially after recognizing that some of the difference is to be attributed to the separation in measurement locations (approximately 60 km at launch, 100 km at altitude).

The second radar being investigated for routine wind profiling is a UHF (915 MHz) radar, now under construction at the Stapleton Airport in Denver. (See Table 1 for main characteristics of this radar.) It is designed to provide considerably lower minimum height (100 m), and better height resolution (90 m) than the 50-MHz VHF radar. It will primarily be used to monitor wind profiles in the atmospheric boundary layer and troposphere; it is not expected to provide data significantly above the tropopause. Unlike the three dipole arrays of the

Platteville radar, it makes use of a single fixed offset parabolic surface (aperture diameter approximately 10 m); this paraboloid is fed by three horn antennas displaced from the focus to obtain three beams, respectively in the zenith, and displaced 15 degrees from the zenith in the north and east directions. It is hoped that this radar, which uses solid state components throughout, will be in routine operation by the end of 1982. Its location will permit easy comparison with radiosonde wind profiles obtained by the National Weather Service twice per day at the same site.

4.2 TEMPERATURE AND HUMIDITY PROFILING

The second major subsystem of the PROFILER is a six-channel microwave radiometer, specifically developed to provide continuous information on temperature and humidity profiles. Four channels in the oxygen absorption band at 5–6 mm wavelengths are used to derive temperature profiles. Corrections to these channels for the effect of any atmospheric water vapour or liquid water are provided by channels at 20.6 GHz and 31.6 GHz. These channels (one on the water vapour line near 22 GHz, and the other in the window between water vapour and oxygen absorptions) serve also to provide information permitting derivation of the water vapour profile, and the vertically integrated liquid water. The radiometer is modelled after the dual-channel radiometer described by Guiraud *et al.* (1979).

The radiometer is housed in an air-conditioned room. The six channels are divided into three pairs of frequencies; each pair is coupled to its own offset parabolic reflector using an orthomode coupler. The six channels have equal antenna beamwidths and receive energy from the zenith via a common low-loss window in the wall of the building and a single flat reflector inclined at 45 degrees immediately outside the window.

In order to calibrate the radiometer, and correct for receiver gain drifts, each channel samples the noise power sequentially from the antenna, a temperature-controlled reference source (at 45°C), and a calibration source maintained at 145°C. Synchronous detection is used to provide a channel output that is proportional to the difference in noise power between the antenna and the reference source. A second synchronous detector, comparing the noise power from the reference source with that from the calibration source, is used to correct the antenna brightness temperature measurements for variations in system gain.

The water vapour and window channels are used to identify the amount of water vapour and liquid water in the beam utilizing adaptive retrieval coefficients (Westwater and Guiraud, 1980) to extend the range of liquid water over which accurate estimates can be made. (The retrieval coefficients are adaptive, in the sense that they are themselves functions of the amounts of water vapour and liquid present.) The derived water vapour and liquid water contents are then used to correct the four oxygen channel brightnesses. These corrected brightness temperatures, plus surface pressure, temperature, and humidity, are used in a statistical algorithm to derive the temperature and humidity profiles. This statistical algorithm is specific to each month of the year and is based on analysis of a multi-year sequence of Denver radiosonde data.

4.3 SURFACE SENSORS

Standard, *in situ* sensors of temperature, pressure, and humidity are used to define the surface values of these parameters, for use in the retrieval of the temperature and humidity profiles from the radiometric data. The values are averaged for 20 min, and stored digitally in the master computer.

4.4 DATA PROCESSING AND MASTER COMPUTER

The VHF radar operates alternately in horizontal and vertical wind-measuring modes, as indicated in Table 1. An FFT algorithm is used to determine the echo spectra, and hence the mean Doppler shift and radial component of velocity, for each range gate on each antenna. The local radar processor then derives the horizontal wind components, u and v, for each height range, for each 130-s interval, and stores these data, together with the vertical profile of vertical wind. In addition, the intensity of the zenithal echo from each range gate is monitored, and used to derive the height of the tropopause (Gage and Greene, 1979). Twenty-minute averages of tropopause height, and of wind components for each 1.5-km height interval, are made available by land-line (upon call) to the master computer at the radiometer site.

The master computer also receives the six-channel radiometric brightness temperatures and retrieves and archives the temperature and humidity profiles as discussed in section 4.2. Additional derived parameters (such as 500 mb pressure height, and profiles of relative humidity or Richardson number) are also potentially available from the computer, in the form of time series, time averages, and time derivatives.

4.5 PERFORMANCE OF THE TEMPERATURE AND HUMIDITY RADIOMETER

Examples of the performance of the PROFILER radiometer system are given in Figs 7 to 12. Simple profiles i.e. those without significant structure, are reproduced with good accuracy (Fig. 7). More complicated profiles, such as Figs 8 and 9 (shown on expanded scales relative to Fig. 7), are considerably smoothed in the mathematical inversion process. This is particularly true of the elevated inversions at heights of about 550 mb in Fig. 8, and 750 mb in Fig. 9. It

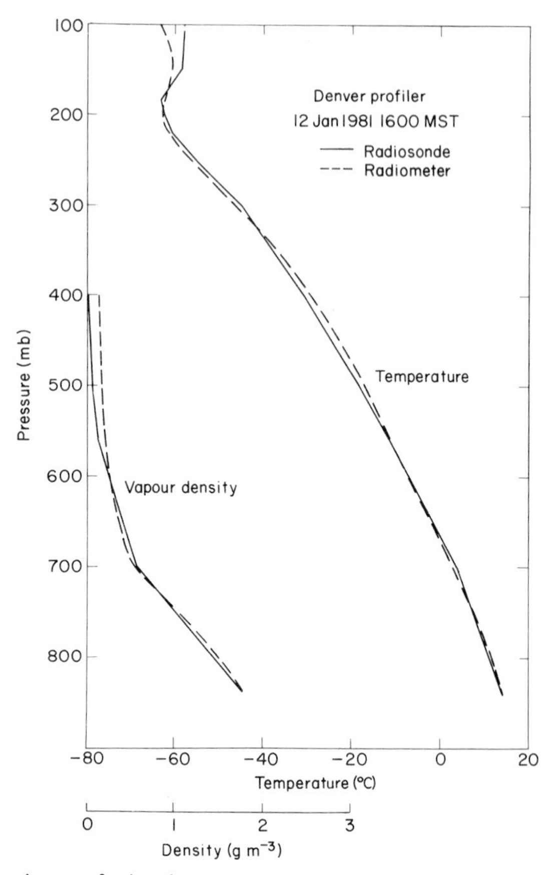

Fig. 7. Comparison of simultaneous, colocated radiometric and radiosonde observations of simple temperature and humidity profiles.

is hoped that the new, high-resolution Stapleton radar will permit identification of the height of significant temperature inversions, and that the use of this information as an additional constraint in the mathematical inversion process will significantly improve the derived profiles. Evidence supporting this expectation is provided in the paper by Westwater and Grody (1980). Preliminary tests of this concept using the Platteville radar estimates of tropopause height (at a location some 60 km from the radiosonde and radiometer site) indicate significant improvement in at least half the cases.

An alternative method of comparing the PROFILER and radiosonde data is in the form of time series. Figure 10 shows a 4-day time series of continuously derived radiometric 700, 500, and 300-mb heights; the dots and vertical error bars denote the seven radiosonde measurements during the period. It will be seen that the 700-mb heights are in excellent agreement; the 500-mb heights agree to

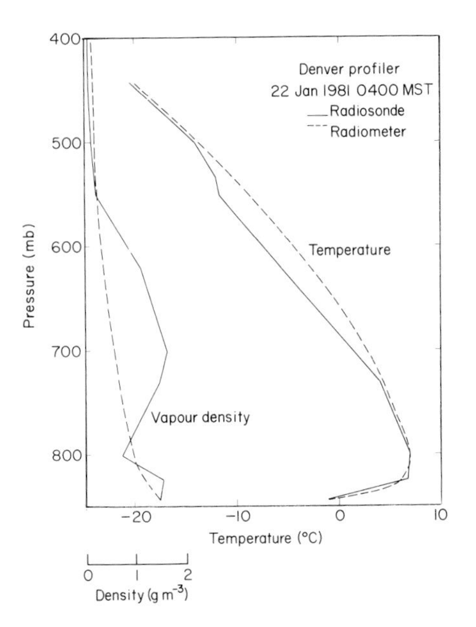

Fig. 8. Ground-based temperature inversion, as recorded radiometrically and with a radiosonde. This case depicts one of the worst agreements so far encountered.

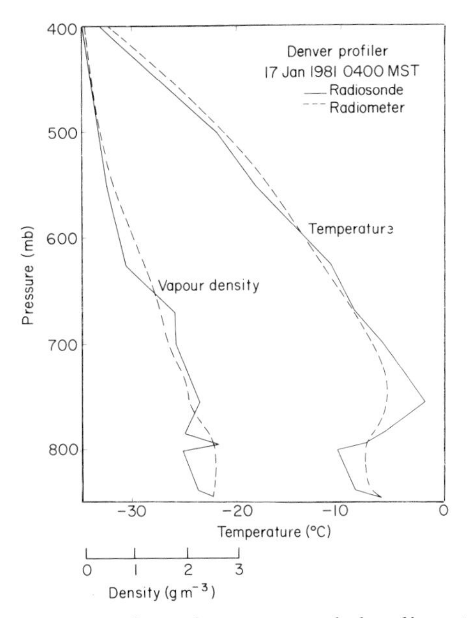

Fig. 9. Elevated temperature inversion, as recorded radiometrically and with a radiosonde.

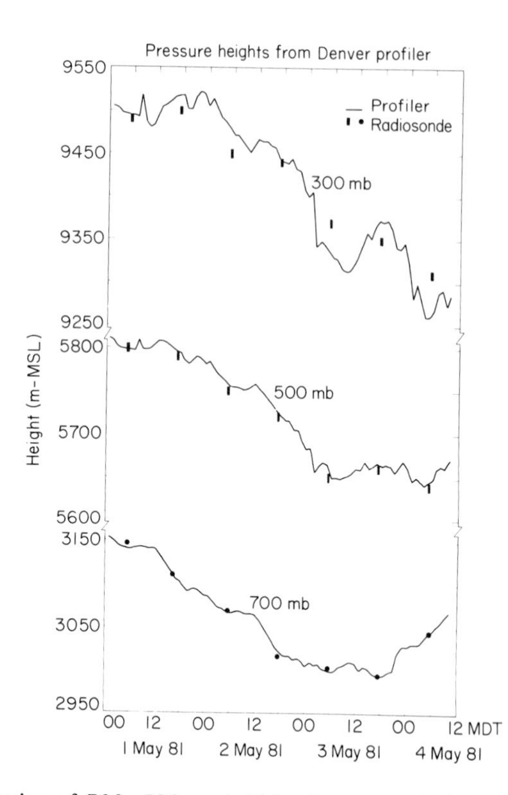

Fig. 10. Time series of 700; 500; and 300-mb pressure heights, as recorded continuously by the radiometer system, and twice daily by radiosondes.

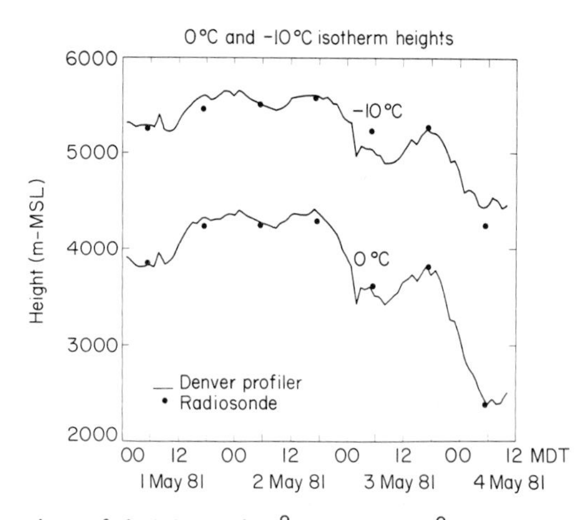

Fig. 11. Time series of heights of 0°C and −10°C temperature levels, as recorded continuously by the radiometer system, and twice daily by radiosondes.

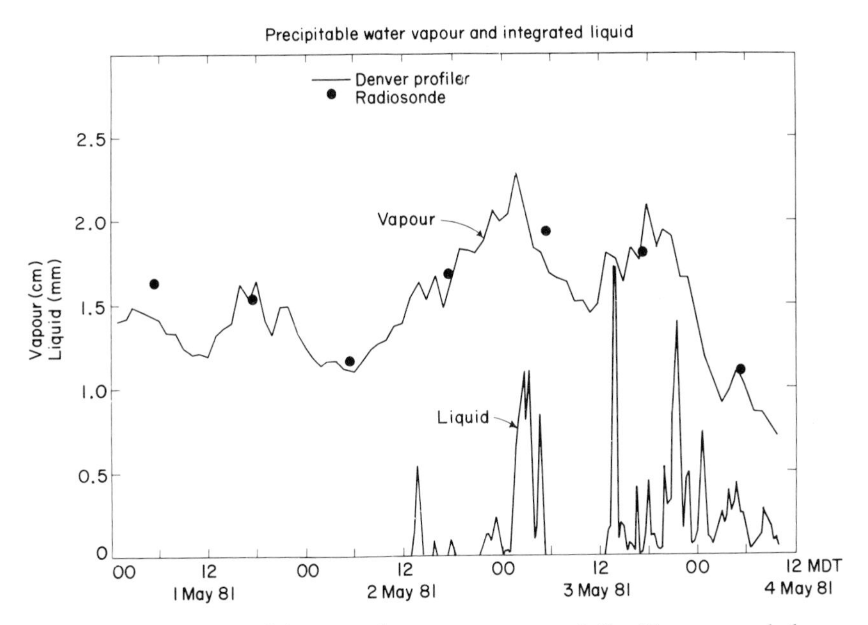

Fig. 12. Time series of integrated water vapour and liquid, as recorded continuously by the radiometer system. The twice-daily radiosonde measurements of total precipitable water vapour are also shown.

within the 10 m rounding used in reporting the radiosonde heights. At 300 mb, the discrepancies are about 25 m r.m.s.; there is some evidence that this number will be reduced by incorporating simultaneous radar measurements of tropopause height in the retrieval process. Figure 11 shows good agreement between the radiometric and radiosonde time series of the height of the 0°C and − 10°C levels. Figure 12 shows time series of zenithally integrated water vapour and integrated liquid water for the same 4-day period, which included periods of significant liquid in clouds and precipitation.

4.6 LIMITATIONS OF THE PROFILER SYSTEM

Preliminary experience with the prototype PROFILER suggests that the desired performance is being achieved. The wind, temperature, and humidity profiles are available essentially continuously to tropopause heights with accuracies and height resolutions which in most cases are consistent with the needs of synoptic scale numerical weather prediction models. The system operates under essentially all weather conditions, though heavy convective rains are expected to degrade the temperature and wind profiles. Equipment reliability has been high, despite the unattended mode of operation. Nevertheless, it should be understood that the temperature profiles, and especially the humidity profiles, have considerably poorer height resolution than the radiosonde, and therefore may not be adequate for some atmospheric research and service purposes.

4.7 SOME POTENTIAL USES OF THE PROFILER

The advent of a radically new geophysical measurement capability typically opens up new research opportunities, followed somewhat later by opportunities for improved or broadened services. The WPL PROFILER, with its ability continuously to measure wind, temperature, and water vapour profiles (plus integrated liquid water) represents such a system. The following offers some thoughts as to its potential uses and impact. These ideas are far from complete, and will undoubtedly be expanded and improved as time progresses, but hopefully they can serve to start others thinking and planning for the future of these systems.

4.7.1 Research opportunities

The main uniqueness of the PROFILER lies in its ability to measure meteorological profiles *continuously in time*, as compared with the intermittently available radiosonde profiles.

This continuous, vertical profiling capability will make it possible to extend studies of the temporal and spatial spectra of atmospheric processes one-to two-decades beyond the one cycle per day, one cycle per 700 km of the existing operational radiosonde network in the United States. This part of the meteorological spectrum includes a wide range of what might be called "significant weather" on the mesoscale (e.g. fronts, thunderstorms, flash floods, chinook winds, land–sea breezes, etc.) whose study and modelling has been handicapped by lack of suitable three-dimensional data. Each of these phenomena requires research with small arrays of continuous profilers. For example, the monitoring of moisture convergence, or of the passage of gravity waves, or of a weather front, would be much better achieved by the continuous profiler than by intermittent radiosondes.

In general, each different atmospheric phenomenon will require PROFILER data sets optimized (especially as to location and spacing) for that specific purpose. Since all meteorological phenomena do not occur in a given geographical area, or on a given scale, it will be desirable that at least some of the research PROFILERS be transportable.

4.7.2 Operational opportunities

Applications to numerical weather prediction

Numerical weather predictions in most areas are currently tied to the 12-h cycle of the upper air (radiosonde) networks. The potential impact on NWP of having vertical profiles available continuously over the network is believed to be large, and may be summarized under five main headings.

1. Improved initialization of NWP models, through improved weather analyses. Since the steadily evolving meteorological fields would be monitored continuously, it should be possible to improve the analysis of the current weather pattern, and hence improve the grid point initialization of the models.
2. Improved initialization of NWP models, through use of time-averaged data and time derivatives. The current upper air input data are basically instantaneous values. A PROFILER array, however, could provide both temporal averages, and time derivatives, as input to the models.
3. More frequent updating of the NWP models. Since the PROFILER data would be available continuously, it would be possible to update the models more frequently, e.g. every 3 or 6 h, instead of once per 12 h.
4. Improved NWP models and parameterizations. At the moment, it is possible to check the models only after they have run 12 hours. Continuously available profiles will permit the careful study of the growth of error in the model relative to the real atmosphere, on time scales of as little as an hour – and hence should permit fine tuning the model, and its parameterizations. (This will be aided by the fact that the initialization of the model should be improved, as a result of factors (1) and (2), above.)
5. In addition to making basic improvements to the NWP products themselves, it should be possible for local weather forecasters to use continuously available local PROFILER data to update and adjust the latest NWP outputs (and associated Model Output Statistics) for their site.

Very-short-range local weather forecasting

It will be some time before numerical weather prediction is used widely for time periods less than 6 to 12 h. The time interval between simple extrapolations (say 1 to 3 h) and the 12-h cycle of current NWP products will therefore involve the use of simple physical models (e.g. for chinook winds, or orographic precipitation) and statistical methods, such as the newly developed GEM methods (Miller, 1979). The continuously available PROFILER data will serve as invaluable input to drive local physical and statistical models.

Calibration of satellite profiling systems

Hitherto, the only comparison standard available for evaluation of satellite sounding data has been the relatively slow, slant profiles obtained by intermittent radiosonde ascents. The continuously available zenithal PROFILER data should offer unique opportunities for comparison with satellite systems such as TIROS-N and GOES/VAS temperature and humidity sounders, and the GOES wind measurement data system. The following four comparison opportunities exist:

1. to use the continuously available PROFILER radiances to provide engineering type calibrations of the performance of the satellite instrumentation and electronics;
2. to use the continuously available PROFILER temperature and humidity profiles to evaluate and refine the retrieval process used in the satellite-derived profiles;
3. to use the continuously available PROFILER wind profiles to evaluate the GOES cloud-image displacement wind measuring system;
4. to use the combination of data from both satellite and ground-based profilers to derive optimum, remotely sensed, profiles.

To a major extent, the satellite and ground-based profiling systems complement each other. Present satellite systems provide good horizontal coverage of the temperature and humidity fields (though the vertical resolution is less than desired, and the accuracy of the measurements degrades near the ground); the satellites fail to provide adequate information on the critically important wind field. On the other hand, a ground-based profiler can provide wind profiles with high accuracy and vertical resolution – but only at one location; its wind, temperature, and humidity profiles are most accurate near the ground, and the radar data can be used to improve the height resolution of the temperature and humidity profiles.

Thus, it seems likely that the optimum upper air profiling system of the future will include some combination of satellites and an array of ground-based profilers to provide three-dimensional fields of winds, temperature, and humidity for numerical (and other) weather prediction purposes.

Acknowledgement

I wish to acknowledge many helpful discussions with NOAA colleagues during the preparation of this chapter, especially M. T. Decker, D. C. Hogg, R. S. Lawrence, R. G. Strauch and E. R. Westwater.

References

Balsley, B. B. (1981). The MST technique – a brief review. *J. atmos. terr. Phys.*, **43**, 495–509.

Brown, E. H. and Hall, F. F. (1978). Advances in atmospheric acoustics. *Rev. Geophys. Space Phys.*, **16**, 47–110.

Clifford, S.F., Ochs, G. R. and Wang, Ting-i (1975). Optical wind sensing by observing the scintillations of a random scene, *Appl. Opt.* **14**, 2844–2850.

Doviak, R.J. and Berger, M. (1980). Turbulence and waves in the optically clear planetary boundary layer resolved by dual-Doppler radars. *Radio Sci.* **15**, 297–317.

Earnshaw, K. B., Wang, Ting-i, Lawrence, R. S. and Greunke, R. G. (1978). A feasibility study of identifying weather by laser forward scattering. *J. appl. Met.* **17**, 1476–1481.

Gage, K. S. and Green, J. L. (1979). Tropopause detection by partial specular reflection using VHF radar. *Science* **203**, 1238–1240.

Guiraud, F. O., Howard, J. and Hogg, D. C. (1979). A dual-channel microwave radiometer for measurement of precipitable water vapor and liquid. *IEEE Trans. Geosci. Elect.* **GE-17** (4), 129–136.

Kaimal, J. C., Baynton, H. W. and Gaynor, J. E. (eds) (1980). Low level inter-comparison experiment, Boulder, Colorado, USA, August–September 1979, Instruments and observing methods, Report no. 3, WMO, Geneva, Switzerland.

Koscielny, A. J., Doviak, R. J. and Rabin, R. (1981). Statistical considerations in the estimation of wind fields from a single Doppler radar and application to prestorm boundary layer observations. (Submitted to *J. appl. Met.*)

Lawrence, T. R., Post, M. J. and Hall, F. F. Jr. (1980). Status of the NOAA ground-based pulsed coherent CO_2 lidar. Topical Meeting on Coherent Laser Radar for Atmospheric Sensing July 5–17, 1980, Aspen, Colorado, sponsored by NOAA and Optical Society of America.

Miller, R. G. (1979). Conditional climatology given the local observation. *Natl. Weath. Dig.* **4**, 2–15.

Ochs, G. R., Clifford, S. F. and Wang, Ting-i (1976). Laser wind sensing: the effects of saturation of scintillation. *Appl. Opt.* **15**, 403–407.

Ochs, G. R., Cartwright, W. D. and Endow, P. S. (1979). Optical system model III for space-averaged wind measurements. NOAA Tech. Memo., ERL WPL-46.

Strauch, R. G. (1979). Application of meteorological Doppler radar for weather surveillance near air terminals. *IEEE Trans. Geosci. Electr.* GE-17 (4), 105–112.

Wang, Ting-i, Earnshaw, K. B. and Lawrence, R. S. (1979). Path-averaged measurements of rain rate and raindrop size distribution using a fast-response optical sensor. *J. appl. Met.* **18**, 654–660.

Wang, Ting-i, Ochs, G. R. and Lawrence, R. S. (1981). Wind measurements by the temporal cross-correlation of the optical scintillations. *Appl. Opt.* **20**, 4073–4081.

Westwater, E. R. and Guiraud, F. O. (1980). Ground-based microwave radio-metric retrieval of precipitable water vapor in the presence of clouds with high liquid content. *Radio Sci.* **15**, 947–958.

Westwater, E. R. and Grody, N. C. (1980). Combined surface- and satellite-based microwave temperature profile retrieval. *J. appl. Met.* **19**, 1438–1444.

Woodman, R. F. and Guillen, A. (1974). Radar observations of winds and turbulence in the stratosphere and mesosphere. *J. atmos. Sci.* **31**, 493–505.

2.2
Nowcasting Applications of Doppler Radar

JAMES W. WILSON and KENNETH E. WILK

1 Introduction

In the past decade there have been major advances in Doppler weather radar and associated improvements in real-time signal processing and colour display. A number of research facilities have been actively testing this new technology to measure remotely air velocities both within storms and optically clear air. As a result of this research a wide variety of Doppler radar applications to improve nowcasting and very-short-range forecasting have been proposed. This chapter reviews the field and presents specific examples of weather phenomena observed by Doppler radar and discusses how this information can be utilized in nowcasting.

Multiple Doppler techniques are proving to be an extremely important method for observing storm kinematics. Three-dimensional fields of air motion are obtained by combining the radial wind components from two or more radars using techniques originally proposed by Lhermitte (1968) and refined by Armijo (1969). Ray *et al.* (1980) demonstrated and compared several alternative schemes for deriving wind fields from multiple Doppler radars. However, multiple Doppler techniques are presently impractical for operational forecasting applications because of the large number of radars required for widespread coverage and the difficulty of combining data from multiple radars in real time. Fortunately, research field programs (Staff, 1979; Doviak *et al.*, 1979; Wilson *et al.*, 1980) are showing that significant nowcasting information can be obtained from a single Doppler radar.

Since a Doppler radar directly measures only the wind component along the beam axis, interpretation of its displays is not straightforward and requires training and experience using mostly pattern recognition techniques. Presently the identification of tornadoes, gust fronts, downbursts and fronts can most easily be accomplished by the forecaster monitoring and interpreting the displays. Other Doppler products such as divergence, vertical and horizontal wind velocities, dangerous wind shear regions, movement vectors, and precipitation rates are easily obtained by computer. Effective Doppler interpretation in real time requires a man–machine mix, with the computer expected to take a greater share of the load as automatic pattern interpretation techniques evolve.

2 Applications

Examples follow of weather phenomena observed by Doppler radar, in clear air (non-precipitating) conditions, widespread precipitation and convective precipitation, from a wide variety of locations in the United States. Doppler *velocity* applications will be emphasized here since the use of radar reflectivity for precipitation surveillance is discussed elsewhere in this volume. Descriptions of the NCAR and NSSL radars and displays are given in Gray *et al.* (1975), Brown and Borgogno (1980), Zahrai (1980), and Hennington (1980).

Generally the displays are in colour; however, they are presented in black and white in this volume, thus reducing the ease of interpretation. The reader is encouraged to look at the colour figures in the Proceedings of the Nowcasting Symposium from which this chapter was condensed. Also some of these same colour pictures are in Wilson *et al.* (1980). Velocity and range ambiguities have been removed, as well as much of the ground clutter. These editing processes are accomplished with interactive computer techniques under development at NCAR (Oye and Carbone, 1981). Real-time signal processing techniques to accomplish some similar results are discussed by Doviak *et al.* (1979).

2.1 WIDESPREAD PRECIPITATION

2.1.1 Wind versus height

Lhermitte and Atlas (1961) first described how a single Doppler radar could be used to measure profiles of wind velocity in widespread precipitation. Their technique, which produces the Velocity–Azimuth Display, or VAD, involves rotation of the antenna while it is directed at a constant elevation angle to record radial velocity of precipitation particles at a fixed range versus azimuth.

Browning and Wexler (1968) presented a thorough analysis of the VAD technique showing how wind velocity, divergence, and deformation of the wind field can be obtained via harmonic analysis. Rabin and Zrnic (1980) have extended the VAD technique to apply to unevenly spaced data. Baynton *et al.* (1977) have described how wind information can be readily obtained visually from the Doppler colour display.

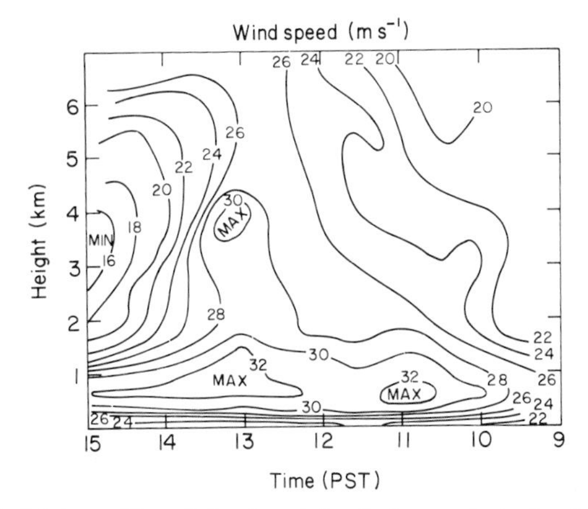

Fig. 1. Time–height profile of horizontal wind speed derived from a Doppler radar, for a 6-hour period on 4 March 1979 at Ocean Shores, Washington. Similar profiles for wind direction, divergence and vertical velocity were also obtained using the VAD analysis technique.

Figure 1 is an example of a time–height profile of wind speed that was obtained by the VAD technique. Similar displays of wind direction, divergence, vertical velocity, and deformation are also possible. These displays can easily be generated in real time with small computers. Such information can be useful in providing the forecaster with a real-time physical understanding of weather events.

Timely wind information could also be utilized by the nowcaster to update local orographic precipitation models similar to that described by Hill and Browning (1981). Vertical wind profiles at each site from a network of Doppler radars would provide numerical models with some of the time and space wind information they require. Testud *et al.* (1980) have shown that the VAD technique can be used to detect the wavelength, amplitude, and direction of propagation of internal gravity waves. Uccellini (1975) has indicated that such waves may be particularly important in triggering severe weather. Wilson *et al.* (1981) have also shown that the VAD technique can be utilized to estimate the average precipitation rate about the radar.

2.1.2 Front location

Sharp wind shifts, which are frequently associated with frontal boundaries, are usually easily identified on the colour velocity display. Thus, many fronts can be precisely located and their movement closely monitored. Figure 2 clearly shows the low-level wind shift associated with a cold front approaching the coast of Washington. In this geographic region accurate frontal locations are rarely known because data are sparse. Note that west of the radar the zero velocity

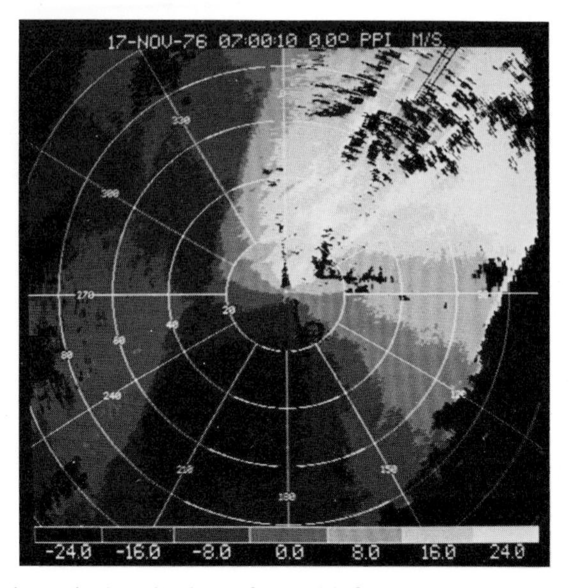

Fig. 2. Doppler velocity display of a cold front approaching Ocean Shores, Washington. The display shows a near-surface sharp wind shift line 30 km west of the radar. Note the sharp bend in the zero contour and close packing of the contours. Winds east of the front about 150 m above the surface are 200° at 24 ms^{-1} and west of the front 280° at 15 ms^{-1}.

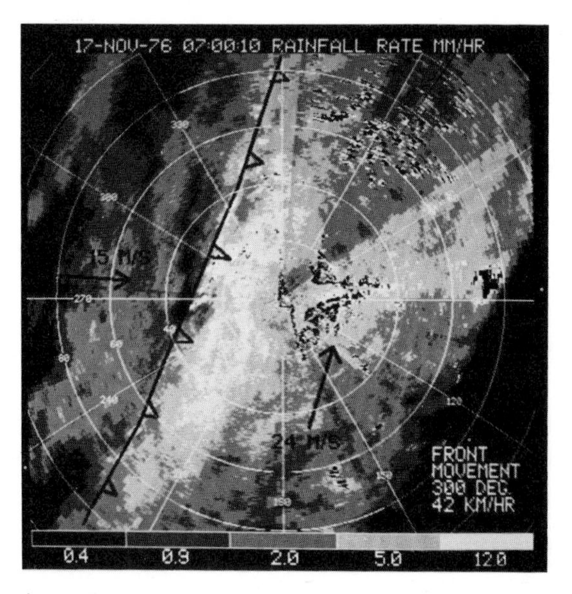

Fig. 3. Derived weather display for the case in Fig. 2 indicating rainfall rate, location of cold front, front movement and near surface wind velocity on either side of the front. This display was generated from Doppler radar data using an interactive computer, colour graphics display system.

band makes a sharp bend from about 290° to 360°. Along the sharp bend and extending to the south-west there is a close packing of the contours. This indicates a sharp change in wind direction from about 200° to 280°. From further inspection we estimate wind speeds of about 24 and 15 ms^{-1} ahead of and behind the front, respectively. Assuming constant advection speed, the forecaster can accurately predict the time of frontal passage and wind velocity after passage.

Figure 3 is an example of a derived-weather display that can be prepared by the meteorologist from the Doppler radar data using interactive graphical techniques. The computer derives the basic rainfall rate display from the reflectivity information. In this case the relationship $Z = 200R^{1.6}$ was used. However, adjustments can be based on known biases or real time rain gauge information (Wilson and Brandes, 1979). The meteorologist may use a lightpen or similar technique to add significant weather information visually extracted from the Doppler velocity display. In this case, the cold front location and wind information were entered in this manner. The movement of the front can be derived by the computer based on a previous position that the meteorologist traced.

2.2 CLEAR AIR

The use of Doppler radar to obtain velocity measurements in the optically clear air was first reported by Lhermitte (1966) and Browning and Atlas (1966). Two principal sources of clear air return emerge from the literature: particulate scatterers and refractive index inhomogeneities. The particulate scatters in most cases appear to be insects. The number, concentration and total radar cross-section of insects have been shown to be sufficient for routine observations using moderately sensitive radars at 3- and 5-cm wavelengths (Browning and Atlas, 1966; Glover *et al.*, 1966; Glover and Hardy, 1966). At radar wavelengths of 10 cm and longer, the dominant scattering appears to result from radio refractive index inhomogeneities in the atmosphere. In the planetary boundary layer, these inhomogeneities result principally from large local gradients of water vapour on the scale of one-half the radar wavelength.

Experience with the NCAR and NSSL 10-cm radars has shown winds can be obtained in the clear air boundary layer out to at least a range of 50 km essentially all the time during the six warmest months. Experience during winter is lacking. The NCAR 5-cm radars frequently observe widespread clear air return in the summer but only very rarely in winter.

Vertical profiles of the horizontal wind can easily be obtained utilizing the VAD technique or from visual inspection. The top of the mixed layer is typically associated with a sharp wind change that is readily evident from the velocity display. Monitoring of the mixed boundary layer height and winds provides much of the information required for air quality nowcasting.

Frequently we observe that a preferred region for thunderstorm initiation is along confluent regions in the clear air. These storms also have a tendency to move down these confluent boundaries while developing into major storms. The confluent regions are visible in the clear air return from the Doppler radar. Monitoring of the location of these boundaries is then an important nowcast tool.

Koscielny *et al.* (1981) have demonstrated that a single Doppler radar can quantitatively map the mesoscale divergence and deformation fields within the cloud-free convective planetary boundary layer. They demonstrated that strong convergence regions preceded cloud and thunderstorm development by 1 to 4 h. These fields, together with vertical profiles of the horizontal wind obtained from the prestorm environment, potentially are powerful tools for forecasting areas of thunderstorm and severe storm development.

The routine detection of clear air echoes in the lowest 2–3 km of the atmosphere during at least the six warmest months of the year points to the utility of meteorological Doppler radar for continuous monitoring of the convective boundary layer winds. It is likely that with a modest increase in sensitivity, state-of-the-art 10-cm radars may be able to measure winds in the optically clear air throughout the entire year. Clear air observations provide much of the information required for short-term air quality forecasting, and to pinpoint covergence areas that initiate deep, moist convection (Harrold and Browning, 1971). Once convective rainfall begins, gust fronts and other turbulent regions outside precipitation regions can be easily identified.

2.3 CONVECTIVE STORMS

The success of Doppler radar for severe storm detection and warning lies in its ability to observe strong velocity gradients in range and in azimuth. Donaldson (1970) showed that data from one Doppler radar are sufficient to identify mesoscale vortices in the wind field. Subsequent work by Burgess (1976) and Brandes (1978) show these signatures are related to the mesocyclone in which tornadoes are frequently embedded and thus can be used for detection and warning.

To verify earlier research on the use of Doppler radar for severe storm warning, an operational test named the Joint Doppler Operational Project (JDOP) was initiated during the spring of 1977 in central Oklahoma. This test reported (Staff, 1979) that cyclonic signatures in the velocity field are precursor indicators of the severity of thunderstorms in Oklahoma and often are reliable indicators of tornadic storms. Numerous regions of cyclonic shear appear in the Doppler display; however, only the strongest and most persistent are likely to produce tornadoes. During JDOP, objective criteria were established for identifying tornado-producing mesocyclones. These criteria involved shear magnitude and temporal/spatial continuity. When the criteria were met they provided an average 20-minute warning before tornado touchdown. Hennington and Burgess (1981) have automated the identification procedure for operational testing.

As indicated previously, we expect for the near future that severe storm identification will be most effectively accomplished by human interpretation of colour Doppler displays. Examples of severe storm features that could be identified by a meteorologist and annotated on a display follow.

2.3.1 Tornadoes

Large tornadoes at close radar range may be associated with very large azimuthal Doppler velocity shear referred to as the tornado vortex signature (Brown *et al.*, 1978). Usually tornadoes are smaller than the pulse volume and are not directly observed. However, they are frequently associated with mesocyclones and a warning can be issued on the detection of the mesocyclone signature. This is particularly true for the larger and more long-lived "Killer" tornadoes. The signature of a mesocyclone is a couplet of closed Doppler velocity contours of opposite sign closely spaced in azimuth (2–15 km between centres). Detection of the mesocyclone is dependent on radar beam width, mesocyclone diameter, strength and range (Zrnic and Doviak, 1975). A typical maximum detection range is 150 km.

Figure 4a is an example of two tornado-producing mesocyclones that were observed in north-central Oklahoma on 2 May 1979. The large mesocyclone, centered at 28° and 60 km, has maximum approaching velocities of 32 ms^{-1} on the west side, and 7 km to the east the maximum receding velocities are 42 ms^{-1} (shear $= 1.1 \times 10^{-2}$ s^{-1}). At the centre of the mesocyclone, there is a tornado vortex signature. Adjoining azimuths indicate speeds changing from -32 to $+42$ ms^{-1}. The other tornado-producing mesocyclone, centered at 5° and 45 km, shows a velocity couplet of -28 and $+21$ ms^{-1} separated by 5 km. Each mesocyclone produced a "hook echo" in the reflectivity display. In some cases, tornado warnings can be issued from the reflectivity display alone, but only after the tornado has formed. Many times hook echoes are not easily identified and hook shapes often exist without tornadoes and mesocyclones.

Figure 4b is the derived precipitation and significant weather display for the case of Fig. 4a. Superimposed on the rainfall rate display are locations of each tornado and its path during the past 30 min as derived from the velocity display. Those echoes in excess of 55 dBz are labelled as possible hail regions. Based on Fig. 4b and stored movement information, the computer can then produce another display (not shown) indicating downstream geographical locations that require warnings.

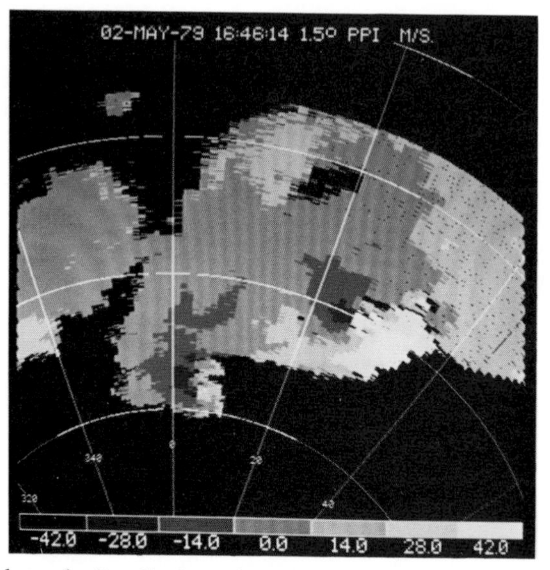

Fig. 4a Doppler velocity display of two tornado-producing storms on 2 May 1979 in north-central Oklahoma. Range rings are at 20 km intervals. (This figure is reproduced in colour in the Frontispiece.)

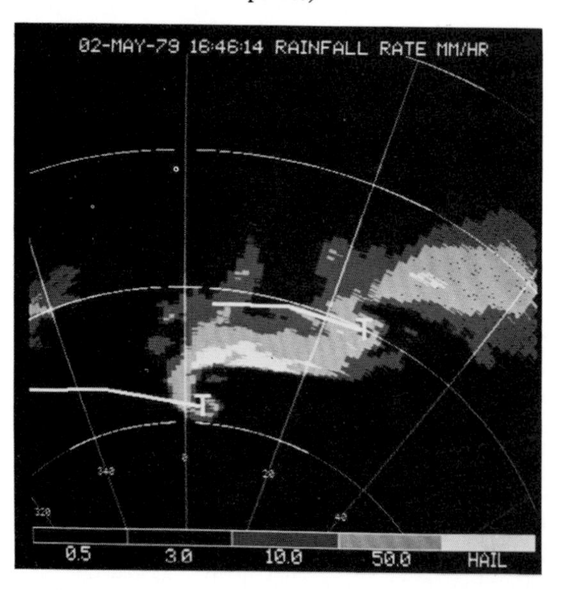

Fig. 4b. Man–machine derived weather display of precipitation and significant weather for the case of Fig. 4a. The grey shades indicate precipitation rate and possible location of hail. The letter T indicates tornado location and the white line their past 30-min path as derived from the Doppler velocity display.

Small tornadoes, without mesocyclones, sometimes form along strong wind shear boundaries, such as gust fronts, downbursts and cold fronts, and are referred to as "gust front tornadoes". These types of tornadoes are more difficult to observe because of their small size and limited vertical extent. However, there is some evidence that they are likely to occur in those regions where the shear is the strongest. One such example is mentioned in the next section.

2.3.2 Gust fronts

Gust fronts are associated with downdraughts descending from the base of cumulonimbus cloud that spread horizontally within several hundred metres of the ground. Many times, this air has a horizontal velocity which is much different from the low-level environmental flow. Consequently, gust fronts generally produce a line or arc of tightly packed velocity contours, indicating a band of strong convergence.

Figure 5a is an example of a gust front that occurred in Illinois on 17 June 1978. Just northwest of the radar, the zero velocity contour is along the 320° radial, indicating a wind direction of 230°. About 30 km from the radar, the zero velocity contour makes a sharp bend to the north, indicating a sudden veering of the wind. From this line of strong shear the gust front is easily traced to the north and northwest of the radar. It cannot be traced south of 300° because either the winds in front of and behind the gust front have similar radial components or the gust front does not exist. Behind the gust front, maximum winds are 32 ms^{-1}.

Figure 5b is the derived weather display for the case of Fig. 5a. Superimposed on the rainfall rate and hail display is the location of the gust front and strong horizontal winds. In this case, the gust front is closely associated with the leading edge of the precipitation. On occasion, the gust front may be 10–20 km in advance of the precipitation. Velocity measurements outside precipitation regions are available routinely during the spring and summer. The scattering mechanisms are normally associated with insects or refractive index fluctuations. Thus, for locating gust fronts, it is often important that the radar have sufficient sensitivity to detect these "clear air" echoes.

For very-short-range forecasting, it is important to monitor gust front locations since they often act as "triggers" for development of new storms or intensification of existing storms.

2.3.3 Downbursts

Downbursts are localized outflows from convective storms that cause damaging winds at the surface. They are similar to gust fronts in that they are associated with downdraughts from a cumulonimbus that spread out into horizontal flow near the ground. Fujita (1981) reports that the width of downburst damage

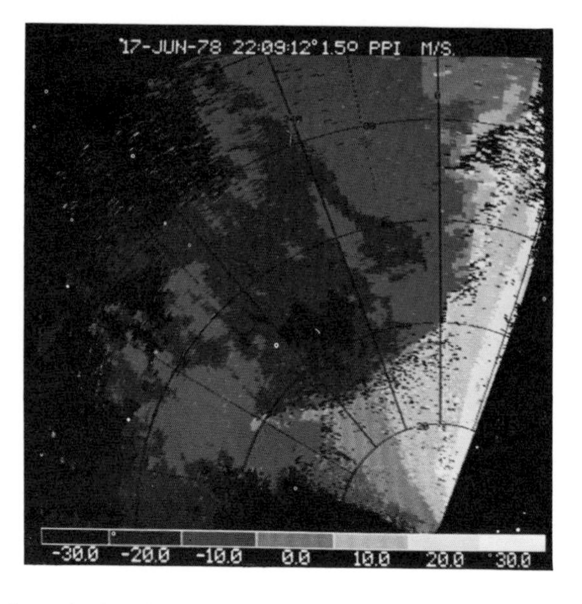

Fig. 5a. Doppler velocity display of a gust front in northern Illinois on 17 June 1978.

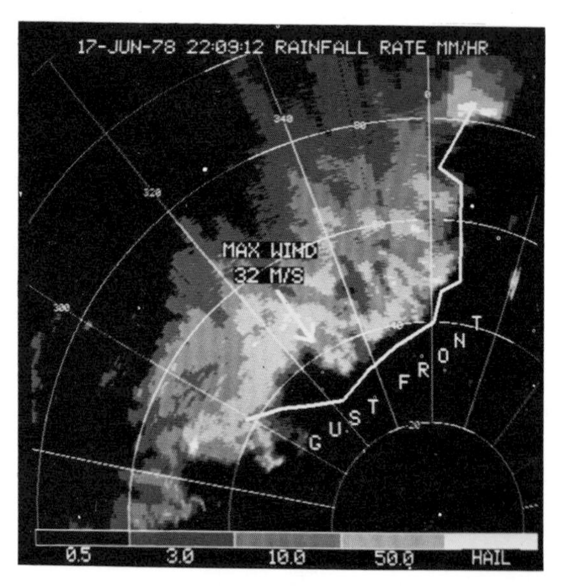

Fig. 5b. Man–machine derived display of precipitation and significant weather for the case of Fig. 5a. The gust front is indicated by the solid white line and the arrow indicates the location and direction of the maximum Doppler velocity wind.

paths are generally less than 20–30 km. Fujita has further defined those with damage paths <4 km as microbursts. On the Doppler display downbursts produce a roughly circular region of closed contours of high radial velocity. This signature is produced by the horizontal outburst of air and not the downdraught. Downbursts may occur within gust fronts or separately.

Figure 6a shows two downbursts within a general area of thunderstorm outflow air. The larger downburst is located at about 155° in azimuth and 42 km in range. The velocity display shows that the air in the downburst is moving away from the radar at 25 ms^{-1}, while only about 5 km to the south and east of this location, the air is moving towards the radar at 12–16 ms^{-1}. A second, but smaller downburst (microburst) of similar magnitude, is observed at 163° and 48 km. The larger downburst rapidly moved towards the east-southeast causing the echo to bulge or form an arc-shaped echo ("bow echo"), as described by Fujita (1979). Ten minutes after the time of Fig. 6a, two small tornadoes developed on the leading edge of the larger downburst in the region of strongest shear. At the time of the tornadoes, the shear at the leading edge of the downburst was $3.5 \times 10^{-2}\ s^{-1}$.

Figure 6b is the derived weather display for this case. Superimposed on the rainfall rate and hail display is the location of the gust front, downburst and strong wind shear regions.

Figure 7 is a wind shear display for this case, indicating the magnitude of the radial shear in the Doppler velocity wind. Not shown but also computed is the azimuthal shear of the Doppler velocity. In this case the shear is computed over a distance of 1.9 km. Radial shear values of $1.5 \times 10^{-2}\ s^{-1}$ are associated with each downburst. This type of display can be quite useful to the meteorologist in assessing the potential for shear-induced tornadoes and in locating mesocyclones. A tangential shear display is better than the radial shear display for mesocyclone identification.

2.4 AIRCRAFT SAFETY

2.4.1 Wind shear

During the past few years, a number of aircraft accidents have occurred in the vicinity of thunderstorms. Fujita and Caracena (1977) and Fujita (1980) have examined six such cases. They concluded that in all cases the aeroplane encountered strong low-level wind shear during take-off or landing. The wind shears were seemingly associated with localized strong downdraughts (or downbursts) from thunderstorms which impacted the surface and then spread horizontally.

In the terminal area, a most important nowcasting application of Doppler radar would be to provide the pilots and controllers information on wind shear

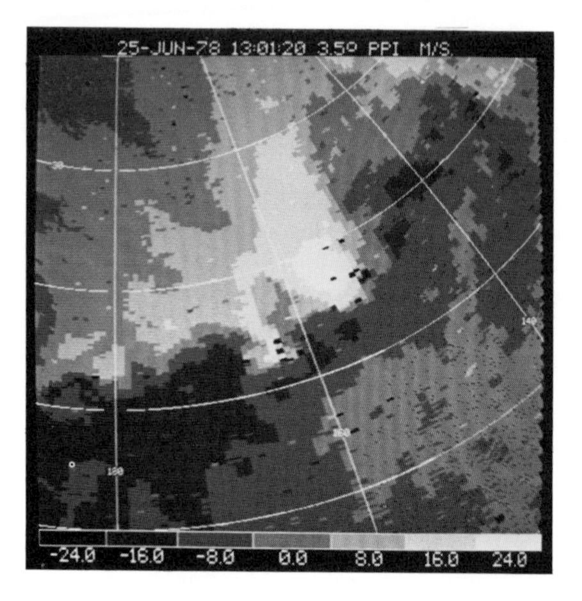

Fig. 6a. Doppler velocity display of two downbursts and a gust front in northern Illinois on 25 June 1978. Range rings are at 10-km intervals.

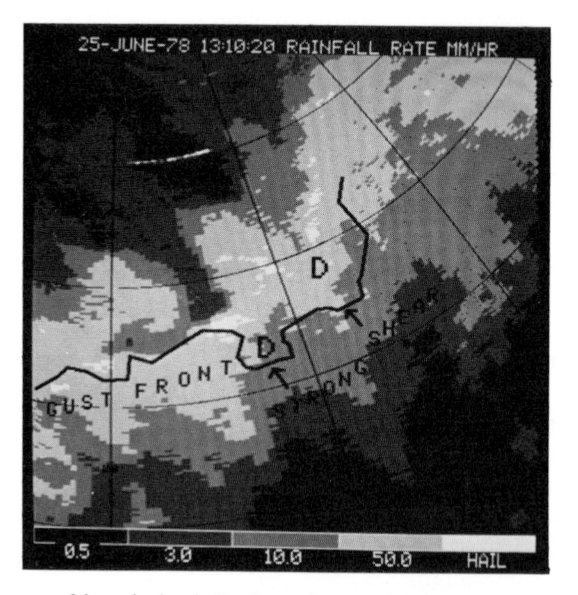

Fig. 6b. Man–machine derived display of precipitation and significant weather for the case of Fig. 6a. The solid black line indicates the location of the gust front, the Ds the location of the downbursts and the arrows the location of strongest wind shear.

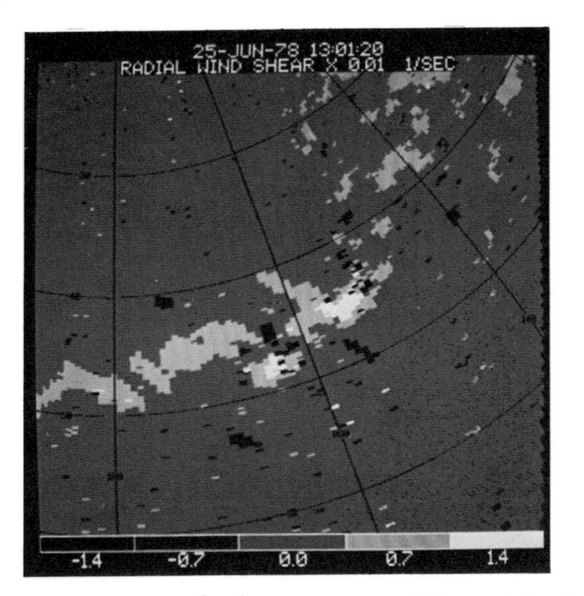

Fig. 7. Wind shear display (10^{-2} s^{-1}) for the case of Fig. 6a. The shear is that of the Doppler velocity computed radially over a 1.9-km distance. Positive values indicate increasing radial velocity with decreasing range.

along approach and departure paths. We can envisage a technique whereby Doppler-measured wind shears are obtained along the glide slopes and that information is given to a computer which contains a numerical simulation model for predicting aircraft performance. Such numerical models have been discussed by McCarthy *et al.* (1980). The wind component of greatest concern to aircraft performance is that along and parallel to the glide slopes. This component can then be measured directly by a Doppler radar located at the airport. The distribution of the vertical wind component is also felt to be important, but only the areal average can be measured by a single radar. For the following applications, we are concerned only with measuring the wind component along the glide slope; thus, it is assumed that the radar is at the airport and that it is scanning along the glide slope. The change in wind speed along the glide slope is obtained by computing the change in the Doppler radial wind with range. The distance over which this computation is made should be commensurate with the spatial scale that is most important to aircraft performance. Numerical models (McCarthy *et al.*, 1980) indicate that aircraft performance is most greatly affected by wind perturbations on a scale of the aircraft phugoid frequency. For jet aircraft similar to the Boeing 727, this is about 2.8 km.

For demonstration purposes, the change in radial wind was computed over a distance of 1.9 km. Pilots are most concerned with change in aircraft airspeed. This information can easily be obtained from a wind shear display (like Fig. 7) by multiplying the shear values by the aircraft speed. Figure 8 is a display of change in airspeed that a plane flying along the glide slope at 140 kt would experience on either approach or take-off. For this demonstration, it is assumed the radar is located at the airport. The hatched region indicates the area where a model utilizing the Doppler wind data as input predicts that a serious hazard would exist to aircraft operations. This example was obtained from a downburst that was approaching the radar on 29 May 1979 in northern Illinois. From this type of display, the approach of the hazard was quite evident and 10–15-min advance warnings could have been provided.

Area-wide displays like Fig. 8 can be continually provided to the tower, approach control, and to enroute air traffic control centres. Potentially hazardous areas would be clearly marked and, when possible, their movement indicated. Flight-specific information of the wind speed and expected airspeed change along a given glide slope can be provided to the pilot. Figure 9 is such an example for a hypothetical approach from 218° at the time of Fig. 8. The wind speed and/or airspeed change data for the glide path can be uplinked directly to the cockpit, thus providing suitable data for pilot-in-command decisions.

2.4.2 Turbulence

Lee (1979) suggests a strong connection between Doppler velocity spectral width measurements in thunderstorms and aircraft measurements of turbulence. During 45 thunderstorm penetrations with research aircraft there were 76 occurrences of moderate or greater turbulence, of which 95% showed Doppler spectral widths of $4\,ms^{-1}$ or greater. The success of this technique depends on the turbulence being isotropic since the aircraft responds principally to vertical gusts and the radar senses horizontal gusts. Displays can be supplied indicating those regions with spectral widths greater than $4\,ms^{-1}$.

3 Conclusions

While we see Doppler radar as a major input to the nowcast and very-short-range forecast, we strongly believe that it must be integrated with data from surface stations, radiosondes, satellites, and numerical weather computer forecasts. The forecaster should review synoptic scale analyses and forecasts and then, by taking into account local topographic influences, prepare a synopsis on how the weather events are likely to develop during his shift. With this background he should be better able to anticipate developments observed with the radar.

For example, we imagine the following scenario for an Oklahoma forecaster on the day shift during May. Upon arrival at work the forecaster reviews the morning radiosonde, synoptic scale analysis, computer forecasts and severe

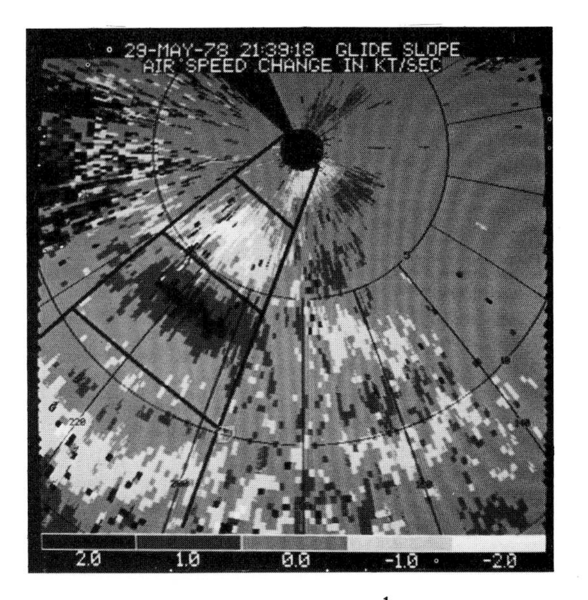

Fig. 8. Display of aircraft airspeed change ($kt\,s^{-1}$) that a plane flying at 140 kt would experience on approach or take-off along the glide slope. For this example it is assumed that the radar is located at the airport. The display is generated from Doppler data collected during a downburst on 29 May 1978 in Illinois. Range rings are at 5-km intervals. The hatched region centred at 6 km toward 218° indicates the area where a serious hazard to aircraft is predicted.

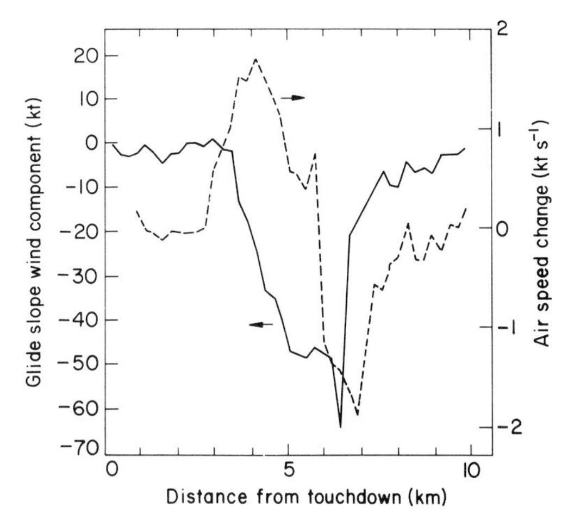

Fig. 9. Example of windspeed (solid line) and aircraft airspeed change (dashed line) versus distance from the airport that can be provided a pilot from a Doppler radar located at the airport. This is the same case as Fig. 8 for a glide slope along the 218° radial.

storm guidance from the National Severe Storm Forecast Center. He determines that the local air mass is too dry for thunderstorms; however, the development of a low in eastern Colorado is forecast and substantial moisture is expected to advect from Texas to central Oklahoma by midafternoon. Also, an area of positive vorticity advection aloft is forecast to arrive about the same time. Thus he concludes that thunderstorm development in the late afternoon is likely, with the possibility that some storms will become severe. As the day progresses he closely monitors the movement of moisture indicated by surface synoptic observations. From the Doppler clear-air returns he locates and tracks the movement of low-level convergence regions (Ogura and Chen, 1977) and areas of dry deep convection (Harrold and Browning, 1971) where thunderstorms are most likely to form. When the low-level moisture is sufficient and/or satellite indicates cloud development, he issues forecasts of thunderstorms in the vicinity of these regions. Also he utilizes the Doppler clear-air return to measure vertical wind shear and surface and upper-air temperature and moisture data to estimate stability in order to anticipate the probable intensity of later thunderstorms (Marwitz, 1972; Fankhauser and Mohr, 1977). Once thunderstorms develop he monitors the Doppler for the issue of severe storm warnings. He examines mid-heights for mesocyclones which are precursors of tornadoes. Lower heights are examined for signatures indicating gust fronts, downbursts, and exceptionally strong shear regions where gust front-type tornadoes may form. Reflectivity information is monitored for precipitation amounts, hail and flash flood potential. This obviously puts a considerable work load on a single person and might actually require two persons, a meteorologist and radar specialist. However, it is important that the meteorologist should spend considerable time examining the Doppler data.

Initially we expect that most warnings will be based on the extrapolation of existing or imminent severe weather conditions identified by the forecaster, and the computer would be utilized to identify downstream locations that need to be warned. We expect that eventually the science will mature to where the "status quo" extrapolation forecast will be modified by anticipating changes. Events that might be used to modify the "status quo" forecast are: storm splitting, storm motion curvature, history of new cell locations, interaction of storm outflow with other existing storms, and cut off of the storm inflow air by its own outflow or the intervention of another storm. We expect that as computers get larger and faster, mesoscale numerical models will begin to play an important role in predicting these changes.

We believe that, if significant advances are to be made in nowcasting, substantial changes will be required in weather service operations. An appropriate mixture of man/machine interaction will be needed to derive Doppler radar products. The nowcaster must be capable of integrating Doppler, satellite, surface and upper-air data and be highly skilled in mesoscale forecasting and knowledgeable about local topographic effects on weather events. Most importantly, efficient rapid dissemination procedures must be available to make available to users easily understood nowcast products.

Acknowledgements

The authors thank Richard Oye of the NCAR Field Observing Facility (FOF), who developed the interactive computer programs for generating the displays, Richard Carbone and John McCarthy for the ideas and stimulation they provided, Edwin Kessler for his review of the paper, and Billie Wheat and Maggie Miller of FOF for typing the manuscript.

The data used in this paper were from three field programs, all at least partially supported by the National Science Foundation. These were the University of Washington CYCLES Project, the University of Chicago NIMROD Project, and the radar portion of the SESAME Project.

References

Amijo, L. (1969). A theory for the determination of wind and precipitation velocities with Doppler radars. *J. atmos. Sci.* **25**, 570–573.

Baynton, H. W., Serafin, R. J., Frush, C. L., Gray, G. R., Hobbs, R. V., Houze, R. A. and Locatelli, J. D. (1977). Real-time wind measurements in extra-tropical cyclones by means of Doppler radar. *J. appl. Met.* **16**, 1022–1028.

Brandes, E. A. (1978). Mesocyclone evolution and tornadic genesis: Some observations. *Mon. Weath. Rev.* **106**, 995–1011.

Brown, R. C. and Borgogno, V. F. (1980). A laboratory radar display system. Preprints 19th Radar Meteorology Conference (Miami) pp. 272–277. American Meteorological Society, Boston Massachusetts.

Brown, R. A., Lemon, L. R. and Burgess, D. W. (1978). Tornado detection by pulsed Doppler radar. *Mon. Weath. Rev.* **106**, 29–38.

Browning, K. A. and Atlas, D. (1966). Velocity characteristics of some clear-air dot angels. *J. atmos. Sci.* **23**, 592–604.

Browning, K. A. and Wexler, R. (1968). A determination of kinematic properties of wind field using Doppler radar. *J. appl. Met.*, **7**, 105–113.

Burgess, D. W. (1976). Single Doppler radar vortex recognition: Part I-Mesocyclone signatures. Preprints, 17th Conference Radar Meteorology Conference (Seattle), pp. 97–103. American Meteorological Society, Boston, Massachusetts.

Donaldson, R. J., Jr. (1970). Vortex signature recognition by a Doppler radar. *J. appl. Met.* **9**, 661–670.

Doviak, R. J., Zrnic, D. S. and Sirmans, D. S. (1979). Doppler weather radar. *Proc. IEEE* **67**, 1522–1553.

Fankhauser, J. C. and Mohr, C. G. (1977). Some correlations between various sounding parameters and hailstorm characteristics in northeast Colorado. Preprints, 10th Conference on Severe Local Storms (Omaha), pp. 218–225. American Meteorological Society, Boston, Massachusetts.

Fujita, T. T. (1979). Objectives, operation, and results of project NIMROD. Preprints, 11th Conference on Severe Local Storms (Kansas City), pp. 259–266. American Meteorological Society Boston, Massachusetts.

Fujita, T. T. (1980). Downbursts and microbursts – An aviation hazard. Preprints, 19th Radar Meteorology Conference (Miami), pp. 102–109. American Meteorological Society, Boston, Massachusetts.

Fujita, T. T. (1981). Tornadoes and downbursts in the context of generalized planetary scales. *J. atmos. Sci.* In press.

Fujita, T. T. and Caracena, F. (1977). An analysis of three weather-related aircraft accidents. *Bull. Am. met. Soc.* **58**, 1164–1181.

Glover, K. M. and Hardy, K. R. (1966). Dot angels: Insects and birds. Preprints, 12th Radar Meteorology Conference (Norman), pp. 264–268. American Meteorological Society, Boston, Massachusetts.

Glover, K. M., Hardy, K. R., Konrad, T. G., Sullivan, W. N. and Michaels, A. S. (1966). Radar observations of insects in free flights. *Science* **154**, 967–972.

Gray, G. R., Serafin, R. J., Atlas, D., Rinehart, R. E. and Boyajian, J. J. (1975). Real-time color Doppler radar display. *Bull. Am. met. Soc.* **56**, 580–588.

Harrold, T. W. and Browning, K. A. (1971). Identification of preferred areas of shower development by means of high power radar. *Q. J. R. met. Soc.* **97**, 330–339.

Hennington, L. (1980). Reducing the effects of Doppler radar ambiguities. Preprints, 19th Radar Meteorology Conference (Miami), pp. 216–218. American Meteorological Society, Boston, Massachusetts.

Hennington, L. P. and Burgess, D. W. (1981). Automatic recognition of mesocyclones from single Doppler radar data. Preprints, 20th Radar Meteorology Conference. American Meteorological Society, Boston, Massachusetts. In press.

Hill, F. F. and Browning, K. A. (1981). The use of climatological and synoptic data for forecasting orographic enhancement of rainfall. *In* "Nowcasting: Mesoscale observations and short-range prediction", pp. 207–212. Proceedings of a Symposium at the IAMAP General Assembly, 25–28 August 1981, Hamburg. European Space Agency, ESA SP-165.

Koscienlny, A. J., Doviak, R. J. and Rabin, R. (1981). Statistical consideration in the estimation of wind fields from single Doppler radar and application to prestorm boundary layer observations. *J. atmos. Sci.* In press.

Lee, J. T. (1979). Potential use of Doppler weather radar for real-time warning of weather hazardous to aircraft. *J. SAFE Assoc.* **9**, 7–11.

Lhermitte, R. M. (1966). Probing air motion by Doppler analysis of radar clear air returns. *J. atmos. Sci.* **23**, 575–591.

Lhermitte, R. M. (1968). New developments in Doppler radar methods. Preprints, 13th Radar Meteorology Conference (Montreal), pp. 14–17. American Meteorological Society, Boston, Massachusetts.

Lhermitte, R. M. and Atlas, D. (1961). Precipitation motion by pulse-Doppler radar. Preprints, 9th Radar Meteorology Conference (Norman) pp. 218–223. American Meteorological Society Boston, Massachusetts.

Marwitz, J. D. (1972). The structure and motion of severe hailstorms: Parts I, II, and III. *J. appl. Met.* **11**, 166–201.

McCarthy, J., Frost, W., Turkel, B. Doviak, R. J., Camp, D. W., Blick, E. F. and Elmore, K. (1980). An airport wind shear detection and warning system using Doppler radar. Preprints, 19th Radar Meteorology Conference (Miami), pp. 135–142. American Meteorological Society, Boston, Massachusetts.

Ogura, Y. and Chen, Y. (1977). A life history of an intense mesoscale convective storm in Oklahoma, *J. atmos. Sci.* **34**, 1458–1461.

Oye, R. and Carbone, R. (1981). Interactive Doppler editing software. Preprints, 20th Radar Meteorology Conference. American Meteorological Society, Boston, Massachusetts. In press.

Rabin, R. and Zrnic, D. (1980). Subsynoptic-scale vertical wind revealed by dual Doppler-radar VAD analysis. *J. atmos. Sci.* **37**, 644–654.

Ray, P. J., Conrad, C. L., Zietler, L., Bumgarner, W. and Serafin, R. (1980). Single- and multiple-Doppler radar observations of tornadic storms. *Mon. Weath. Rev.* **108**, 1607–1625.

Staff, NSSL, AFGL, NEW and AWS, (1979). Final Report on the Joint Doppler Operational Project (JDOP) 1976–1978. NOAA Tech. Memo. ERL NSSL-86. Norman, Okla, 84 pp.

Testud, J. Breger, G., Amayenc, P., Chong, M., Nutten, B. and Sauvaget, A. (1980). A Doppler radar observation of a cold front: Three dimensional air circulation, related precipitation system and associated wave-like motions. *J. atmos. Sci.* **37**, 78–98.

Uccellini, L. W. (1975). A case study of apparent gravity wave initiation of severe convective storms. *Mon. Weath. Rev.* **103**, 497–513.

Wilson, J. W. and Brandes, E. A. (1979). Radar measurement of rainfall – A summary. *Bull. Am. met. Soc.* **60**, 1048–1058.

Wilson, J., Carbone, R., Baynton, H. and Serafin, R. (1980). Operational Application of Meteorological Doppler Radar. *Bull. Am. met. Soc.* **61**, 1154–1168.

Wilson, J. Carbone, R. and Ramsay, B. (1981). Precipitation and precipitation efficiencies derived from single Doppler radar. Preprints, 20th Radar Meteorology Conference. American Meteorological Society, Boston, Massachusetts. In press.

Zahrai, F. A. (1980). Real-time Doppler radar data processing and display. Preprints, 19th Radar Meteorology Conference (Miami), AMS Boston pp. 211–215.

Zrnic, D. C. and Doviak, R. J. (1975). Velocity spectra of vortices scanned with a pulse-Doppler radar. *J. appl. met.* **14**, 1531–1539.

2.3
Mesoscale Observations from a Polar Orbiting Satellite Vertical Sounder

G. A. M. KELLY, B. W. FORGAN, P. E. POWERS and J. F. LE MARSHALL

1 Introduction

The TIROS-N NOAA A-G series of satellites began operating in October 1978. These satellites are high-quality observation platforms supplying large amounts of data describing the Earth's surface and enveloping atmosphere. The TIROS-N Operational Vertical Sounder (TOVS) instrument packages on board all these satellites allow vertical temperature and moisture structures (soundings) to be calculated between the surface and the stratopause. Ideally two satellites are operational and are in sun-synchronous polar orbit at any one time. This implies a full global coverage by these satellites every 6 hours.

The TOVS package consists of three radiance measuring instruments: the High-resolution Infrared Radiation Sounder (HIRS), sampling at 20 infrared frequencies, the Microwave Sounding Unit (MSU), sampling at 4 microwave frequencies, and the Stratospheric Sounding Unit (SSU), sampling at three additional infrared frequencies. Typical scan patterns of the HIRS and MSU instruments across the suborbital track are shown in Fig. 1. The HIRS instrument resolves a mean area 30 km in diameter at the sub-satellite point, whereas the MSU resolves a circular area of 110 km in diameter. There are 56 fields of view (fov) within each scan line width of approximately 2250 km for the HIRS instrument. Only 11 fovs are completed in every MSU scan line in the time the HIRS instrument takes for 5 scan lines or 280 fovs. For both instruments, the fov "footprints" become elliptical and enlarge as the fov location is removed further from the sub-satellite point.

The decision to receive and process these data locally was based on operational forecasting requirements. Meteorological data of up to 60 km resolution are available within one hour of a satellite pass after TOVS processing. Optional interactive processing of the locally readout data gives the ability to delete or correct erroneous data and display a variety of analysed fields. Although this

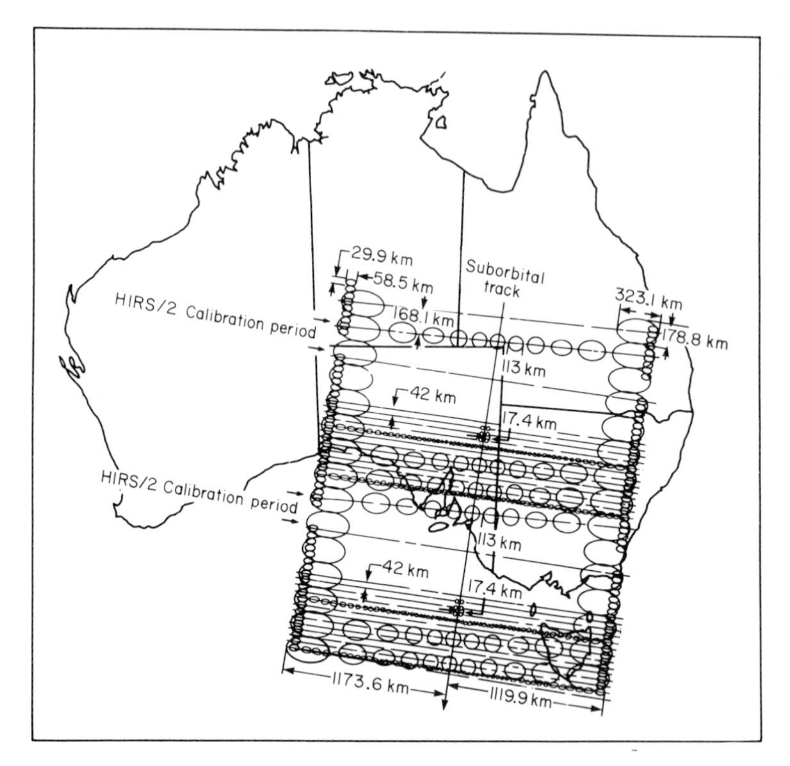

Fig. 1. Typical TOVS orbit showing the resolution of the HIRS and MSU instruments. The orbital direction is from the top right to the bottom left of the figure.

procedure was not required in the case presented below, it is used occasionally in operations. These advantages, combined with the ability to display derived fields on request at high spatial resolution, make the local retrieval system a powerful tool in the hands of the weather forecaster.

In November and December 1980 a special observing program was mounted to study summer-time cold fronts over south-eastern Australia. As part of this study, two-hourly radiosonde flights were undertaken at Mt Gambier (WMO station number 94821). These measurements, together with other elements of the data base, will be used to illustrate the ability of the TOVS data to detect a mesoscale cold pool.

2 The TOVS processing system

Raw TOVS data from the TIROS-N Information Processor (TIP) are acquired directly by reception of the VHF beacon Direct Sounder Broadcast from the TIROS-N satellites. These TIP data contain information from the TOVS package as well as the ARGOS data collection and location system, the Solar Environment Monitor (SEM) and other environmental packages.

The organization of the recorded output from the TOVS data-collection system into regular file structures suitable for scientific processing is performed by means of the preprocessing programs. Further corrections and calibration of the radiance data are carried out and the resulting data stored in suitable forms for display and retrieval processing. Three additional files are used: regression coefficient files for the HIRS and MSU limb corrections, together with a topography file for ascertaining topography type and elevation (resolution $1^\circ \times 1^\circ$ latitude–longitude).

The low resolution MSU channel data are mapped onto the high-resolution HIRS data by a distance weighted interpolation scheme. From these mapped fields and ancilliary data two files are created. The first file, called the IMAGE file, is formatted to facilitate the display of horizontal fields for any recorded parameter. The second file, or SOUNDER file, is arranged to provide vertical profiles of radiance measurements for each HIRS location. It is from this sounding file that the retrievals are calculated.

The initial function of the retrieval section is to provide sets of atmospheric soundings which have been filtered spatially and corrected to remove the effects of cloud on the observed radiances. This cloud contamination is one of the main problems in supplying representative soundings. While the original resolution of the sounding file is one spot per 30 km, atmospheric temperature and moisture profiles every 60 km are produced via the retrieval process. The retrieval scheme used is a multiple linear regression technique determined from the correlation of a large number of coincident radiosonde profiles and satellite radiance measurements. It was first described by Smith *et al.* (1970) for use with Nimbus-3 SIRS observations. For each retrieval spot the eight surrounding radiances are used to provide representative uncontaminated soundings for a 3600 km^2 area.

Further checks on the temperature profiles are performed to remove inconsistent ones, thereby reducing the number of retrievals. This redefined set of retrievals is supplemented by including profiles derived from the MSU data alone where large gaps in the cloud-corrected retrievals have occurred because of cloud contamination. This enhanced set of retrievals is then re-examined for inconsistent or redundant data. Finally, the retrieved profiles are used to calculate thermal winds (shears) by analysis of the geopotential thickness fields. Although any products of the sounding or retrieval process may be displayed, the normal display data consist of analysed fields of temperature, geopotential thicknesses and wind vectors.

The resultant temperature soundings from the retrieval process compare well with the radiosonde data. Figure 2 shows the r.m.s. differences between the TOVS and Australian radiosonde soundings. The largest differences are near the

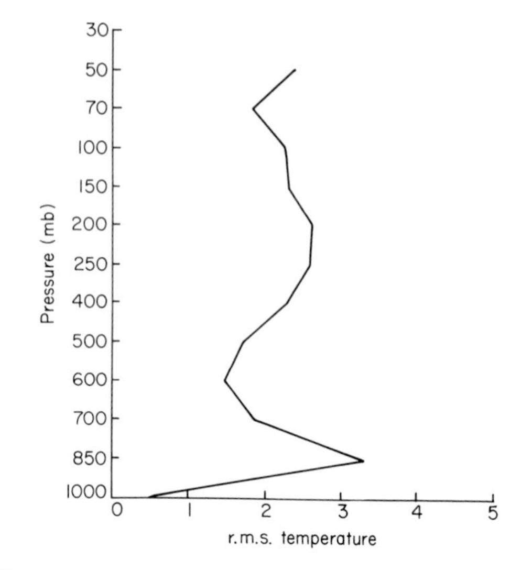

Fig. 2. r.m.s. differences between colocated TOVS and radiosonde measurements. The number of samples used was 289, with colocation assumed if measurements were within 1 hour and 100 km.

surface and in the tropopause region. These are caused by fine-scale vertical structures below the resolving power of the TOVS. However, the r.m.s. temperature differences for the mid-troposphere near 500 mb are not much more than 1 K.

3 Analysis using TOVS data

The impact of mesoscale observations from the TIROS-N series of satellites on broadscale analysis is illustrated below. An example has been drawn from the special cold front observing period with pertinent broadscale (National Meteorological Analysis Centre, NMAC) and mesoscale (TOVS) analyses being contrasted.

3.1 BROADSCALE ANALYSIS

Broadscale numerical analysis in NMAC (Seaman, 1977) relies on initial specification of mean sea level pressure (MSLP) and 1000/500 mb thickness by manual analysis methods. These methods use the conventional observation network, low resolution (500 km) satellite-derived temperature profiles where available, together with cloud picture interpretation (Kelly, 1978; Guymer, 1978), to extend the data base into areas of no data. The final numerical analysis is performed on a 24 by 39 grid of 220 km resolution.

Examples of the fields resulting from such analyses in the area of the orbiting satellite pass on 12 December 1980 at 23 GMT are shown in Fig. 3. Shown is a weak cold trough which developed south of Australia. Cloud formations resembling a small-scale vortex with an associated cloud band are shown in Fig. 4. The development and evolution of similar cloud formations have been discussed by Anderson *et al.* (1974) and are often described as "comma clouds" (which are inverted in the Southern Hemisphere). Figure 4 shows that at 2330 GMT the vortex centre was located near 40°S 134°E and the associated cloud band showed frontal characteristics with stratocumulus clearing immediately ahead of it. In this particular case no low resolution TOVS data were available in the area of the vortex at the time of the operational analysis.

3.2 MESOSCALE ANALYSIS

The analysis has been repeated using the mesoscale data base afforded by the TOVS instrument. The spatial resolution of the TOVS temperature retrieval is displayed in Fig. 5. The sounding density is reduced in regions of extensive cloudiness and during periods of HIRS calibration. Temperature, moisture and geopotential thicknesses are calculated and then analysed on a one-third HIRS resolution (approximately 60 km) grid at each level using the successive correction method (Cressman, 1959).

Figure 6 displays two temperature field analyses together with conventional wind and temperature observations at two mandatory pressure levels: 500 mb and 250 mb. At 500 mb a cold pool is clearly evident at 38°S 130°E and at 250 mb a warm pool is located about 3 degrees to the northwest. In the previously shown broadscale analysis at 500 mb (Fig. 3(a)) the cold pool was depicted as a trough. This clearly shows the value of the TOVS high-resolution measurements for analysis. Figure 7 shows geopotential height fields derived from the TOVS thicknesses and NMAC surface pressure analyses. Again the TOVS data show the cold pool to be well developed throughout the middle troposphere.

Another important quantity derived from TOVS data is the geostrophic wind. The high resolution TOVS data provide good quality horizontal thermal temperature gradients from which geostrophic winds can be calculated. Geostrophic winds derived from TOVS are shown in Fig. 8 together with the conventional wind values derived from radiosondes. The spatial consistency of the TOVS wind field, combined with the good agreement with pilot balloon winds, points to the utility of the derived fields.

4 Comparison with radiosondes

In order to compare radiosonde measurements with profiles derived from TOVS it is necessary to consider what each instrument is sensing. The radiosonde

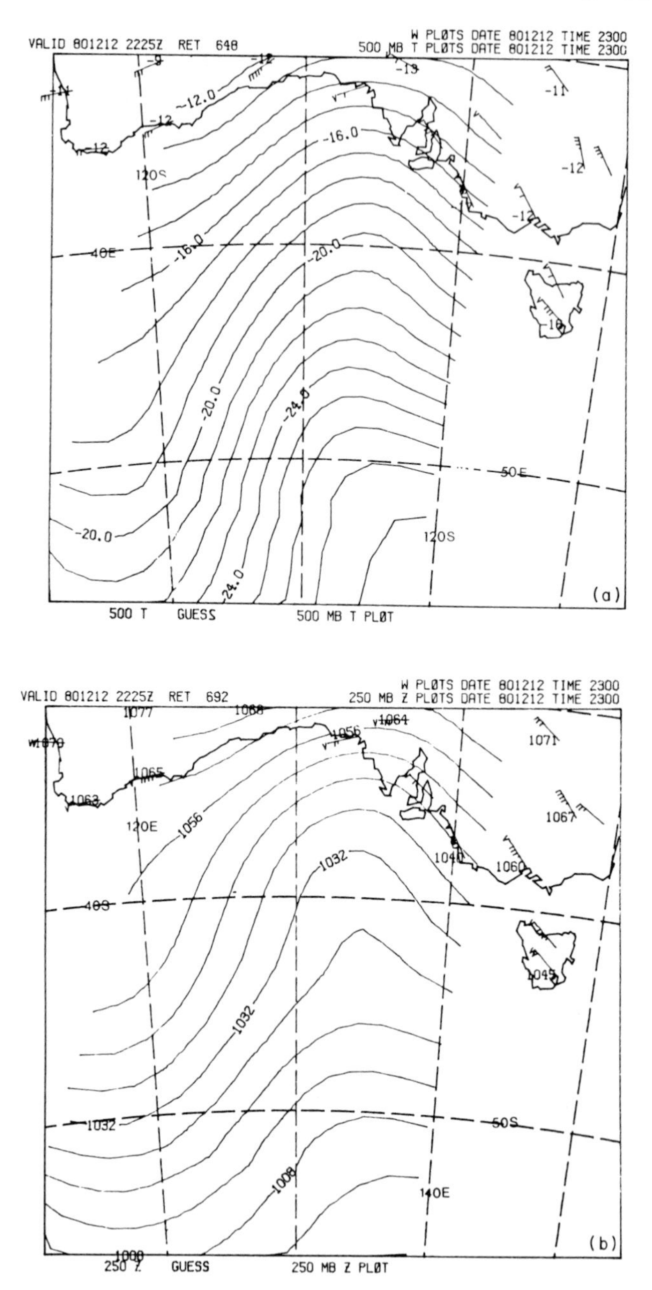

Fig. 3. NMAC analyses for 12 December, 1980 at 23 GMT. 500 mb temperature (°C) is shown in (a) and 250 mb geopotential heights (dm) are shown in (b).

Fig. 4. Visual channel cloud picture from the geostationary satellite GMS for 12 December 1980 at 2330 GMT.

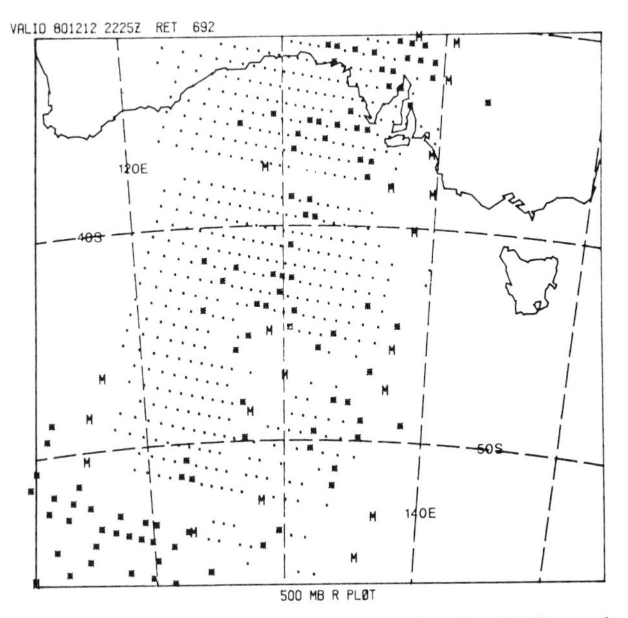

Fig. 5. A plot of TOVS retrieval type and location for 12 December 1980 at 23 GMT. The three types are as follows: clear (•), cloudy (*) and microwave only (M).

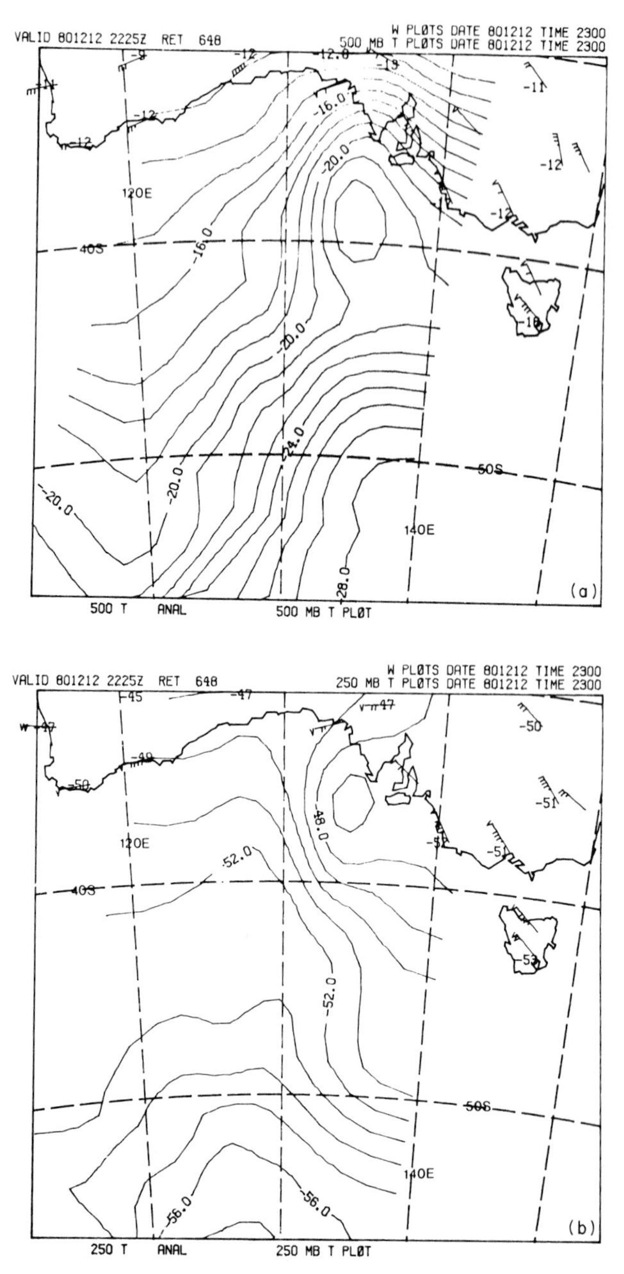

Fig. 6. Temperature analyses using TOVS data as shown in Fig. 5 together with conventional temperature and wind observations: (a) 500 mb (°C), (b) 250 mb (°C).

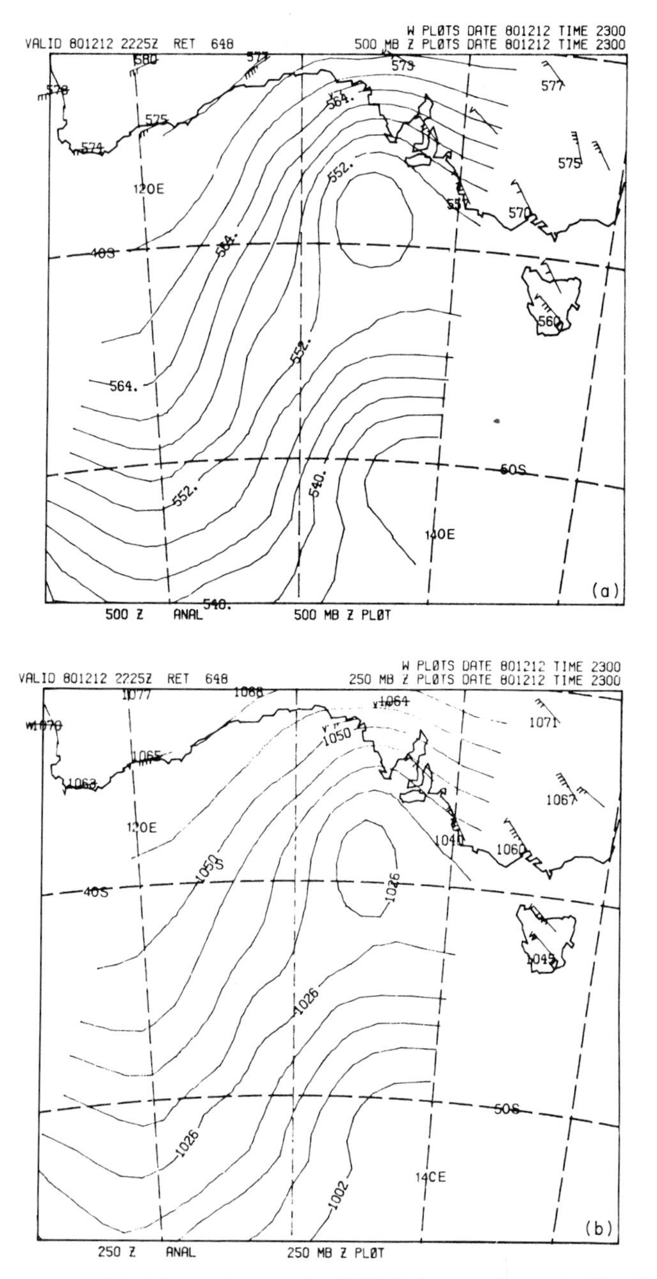

Fig. 7. Geopotential height analyses using TOVS data as shown in Fig. 5 together with conventional geopotential heights and wind observations: (a) 500 mb (dm), (b) 250 mb (dm).

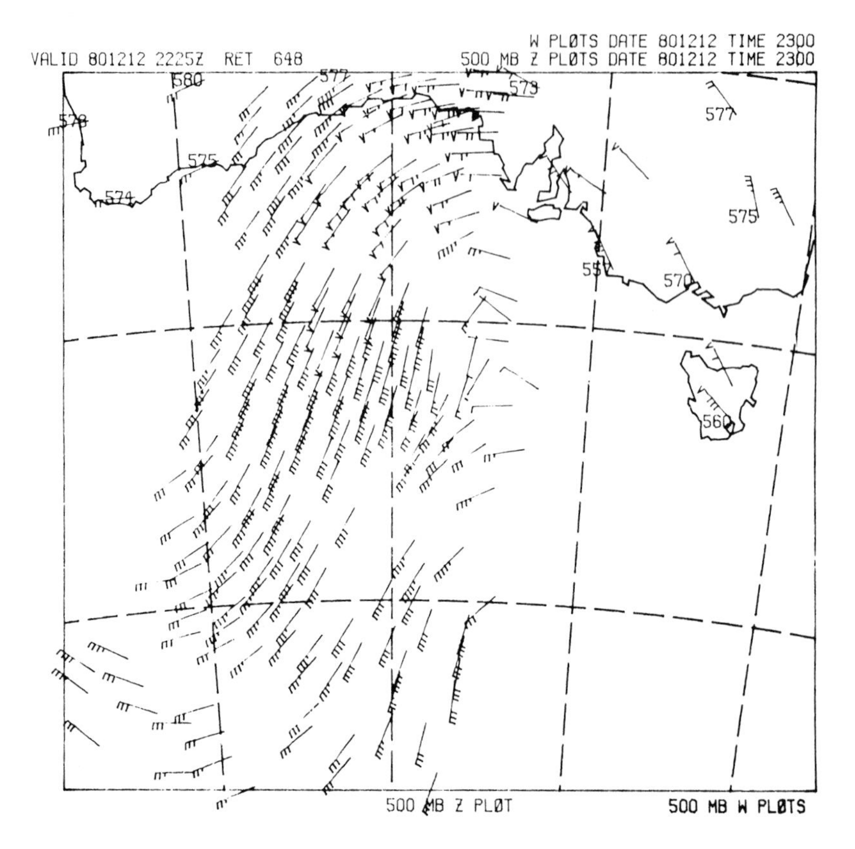

Fig. 8. Geostrophic winds at 500 mb derived from TOVS data as shown in Fig. 5, and conventional geopotential wind and height observations.

provides a point measurement of temperature at various levels in the atmosphere. On the other hand the TOVS radiance measurements sense the temperature of a layer of the atmosphere averaged over 30 km^2 in the horizontal (Smith *et al.*, 1979).

Figure 9 shows four comparisons between radiosonde and TOVS profiles for the study period. In general there is good agreement both in the middle troposphere and near the tropopause. The radiosonde has a finer vertical resolution and therefore shows much more detail near the surface and tropopause. The difference in vertical resolution is emphasized in Fig. 9(c) where a strong surface inversion exists due to extensive low cloud. This is displayed in the radiosonde trace but is missed by the TOVS profile. Also given in Fig. 9 are the HIRS–MSU temperature differences, which are important for quality control of TOVS-derived profiles.

During the special observing period at Mt Gambier (38°S 141°E) two-hourly radiosonde flights were conducted, and TOVS passes occurred at 6-hourly intervals. The cold pool first moved towards the northeast during this period and passed close to the station before moving to the southeast. Figure 10(a) shows the time section derived from the radiosonde flights during this period. Assuming that little change occurred during this period, the section gives an effective cross-section of the cold pool and associated upper jet structure. This section can be compared with a cross-section taken through the cold pool across the TOVS orbital pass (Fig. 10(b)). In both sections the cold pool is mid-tropospheric, lying between the 32°C and 36°C isentropic surfaces. The jet stream structure is also similar in both sections with the leading jet core speeds being in excess of $50\,m\,s^{-1}$.

5 Concluding comments

Satellite-derived temperature soundings are vital for mesoscale analysis in the Southern Hemisphere particularly over ocean regions. The comparisons with radiosondes give confidence in the reliability of the data. It is now possible to analyse both synoptic and mesoscale systems, and hence provide both forecasters and numerical prognosis models with a more accurate analysis. These TOVS data have already been used to provide analyses for a movable fine mesh model (Gauntlett, 1981). There still remain some significant problems, including the effect of cloud on observed radiances and the need for the reduction of large r.m.s. errors near the surface.

Acknowledgements

The authors wish to acknowledge the significant help and encouragement provided by Dr W. L. Smith and his staff at the University of Wisconsin.

References

Anderson, R. K., Ashman, J. P., Bittner, F., Farr, G. R., Ferguson, E. W., Oliver, V. G., Smith, A. H., Purdoin, J. F. W. and Skidmore, R. W. (1974). Application of meteorological satellite data in analysis and forecasting. ESSA Technical Report, NESC 51.

Cressman, G. P. (1959). An operational objective analysis. *Mon. Weath. Rev.* **87**, 367–374.

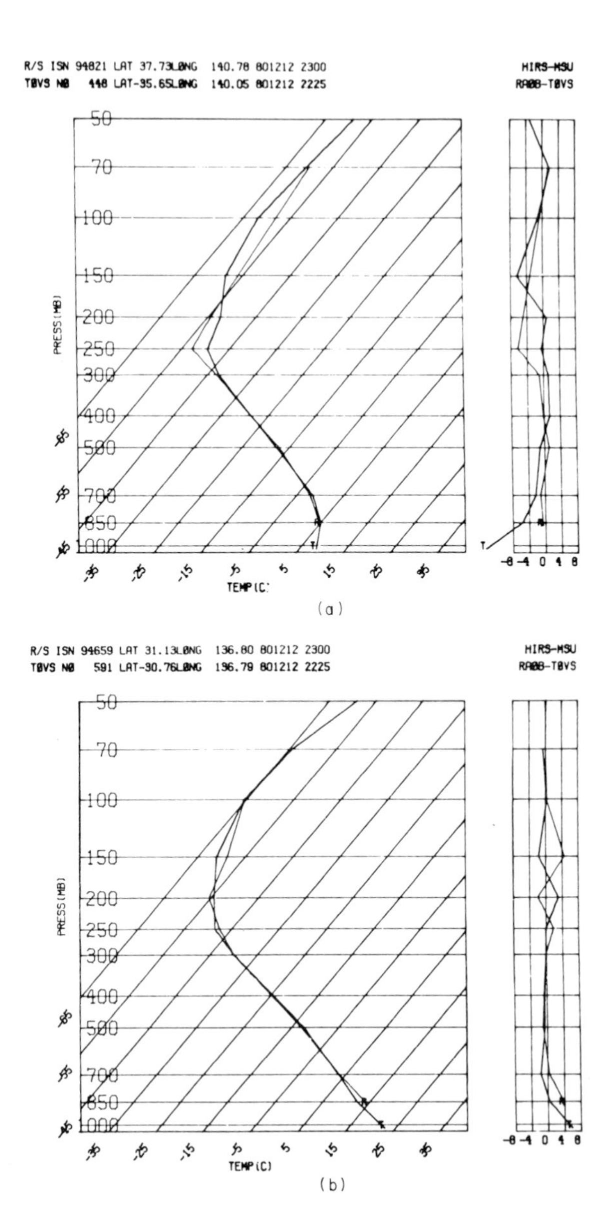
R/S ISN 94821 LAT 37.73LØNG 140.78 801212 2300
TØVS NØ 448 LAT-35.65LØNG 140.05 801212 2225
HIRS-MSU
RAØB-TØVS
PRESS(MB)
TEMP(C)
(a)
R/S ISN 94659 LAT 31.13LØNG 136.80 801212 2300
TØVS NØ 591 LAT-30.76LØNG 136.79 801212 2225
HIRS-MSU
RAØB-TØVS
PRESS(MB)
TEMP(C)
(b)

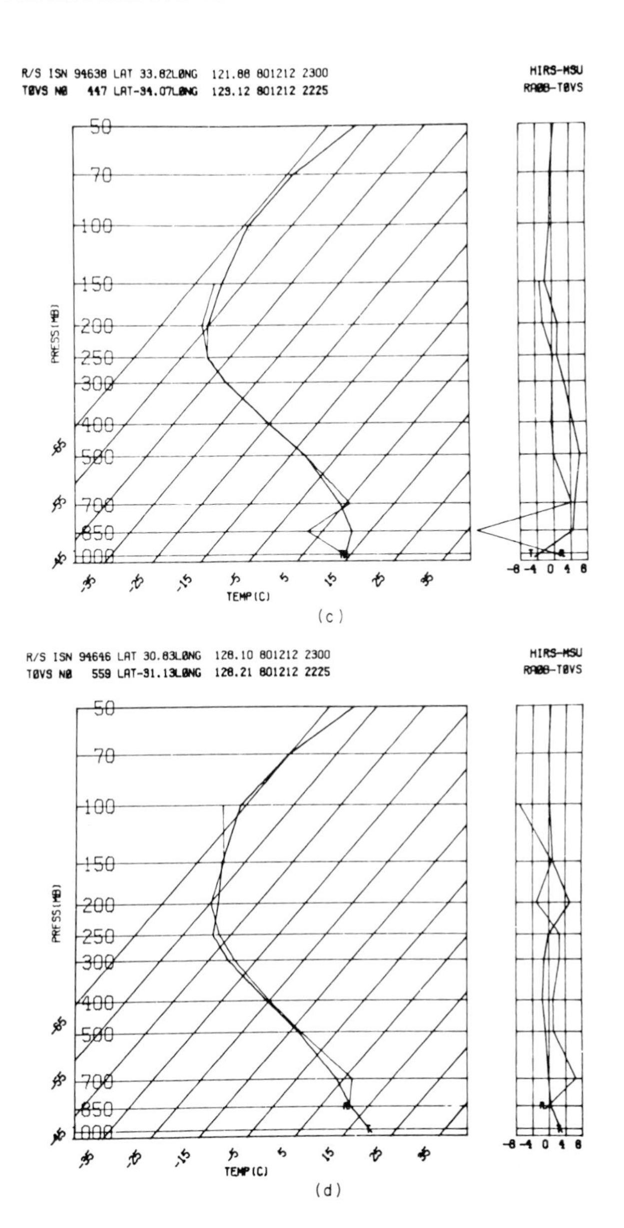

Fig. 9. Aerological diagrams (Skew T/Log P) of TOVS and conventional radiosonde soundings for four locations: (a) Mt Gambier (94821), (b) Forrest (94659), (c) Esperance (94638), (d) Woomera (94646. Left side: Skew T/Log P, TOVS (T) radiosonde (R). Right side: difference plots, TOVS-radiosonde (R), TOVS HIRS-TOVS MSU (T).

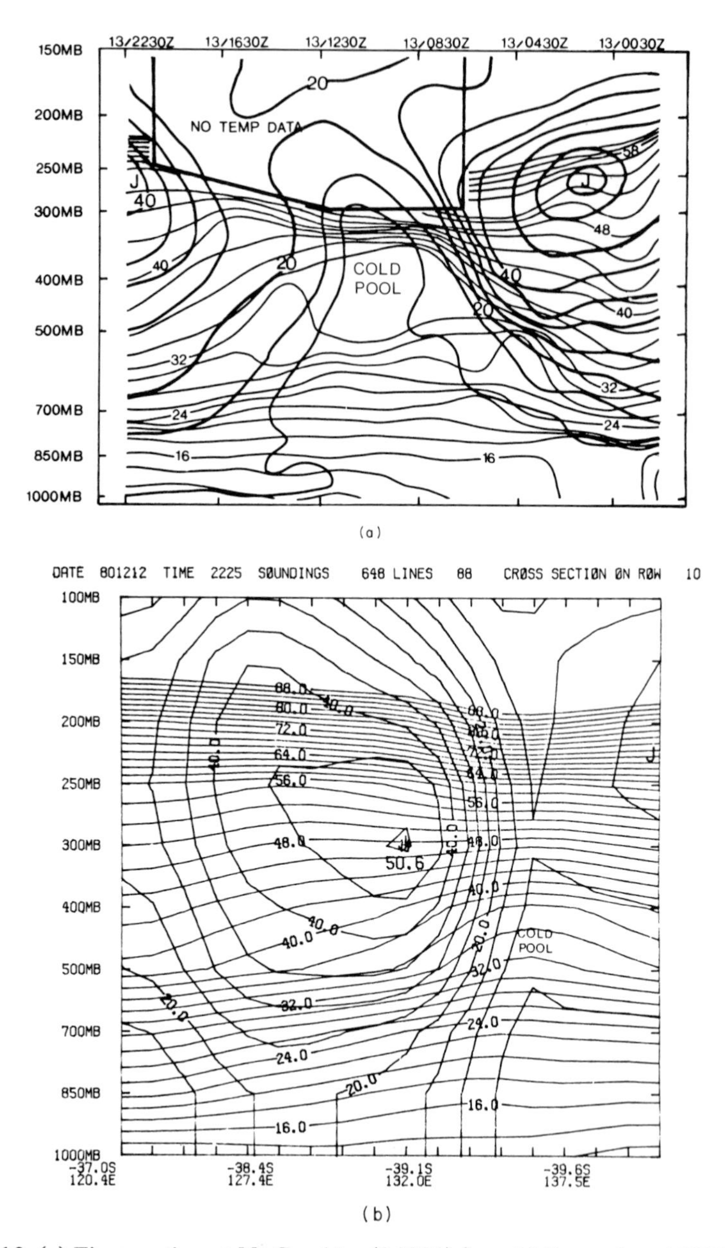

Fig. 10. (a) Time section at Mt Gambier (94821) from 12 December at 2030 GMT to 13 December 1980 at 2230 GMT, showing potential temperature (°C) and wind (m s^{-1}). (b) Cross-section across the TOVS orbit showing potential temperature (°C) and wind (m s^{-1}).

Gauntlett, D. J. (1981). The numerical simulation of intense frontal discontinuities over south-eastern Australia. In "Nowcasting: Mesoscale observations and short-range predictions" pp. 271–275. Proceedings of a symposium at the IAMAP General Assembly, 25–28 August 1981, Hamburg. European Space Agency ESA SP-165.

Guymer, L. B. (1978). Operational application of satellite imagery to synoptic analysis in the Southern Hemisphere. Technical Report, 29, Bureau of Meteorology, Melbourne, Australia.

Kelly, G. A. M. (1978). Interpretation of satellite cloud mosaics for Southern Hemisphere analysis and reference level specification. *Mon. Weath. Rev.* **106**, 870–889.

Seaman, R. S., Falconer, R. L. and Brown, J. (1977). Application of a variational blending technique to numerical analysis in Australian region. *Aust. met. Mag.* **25**, 3–20.

Smith, W. L., Woolf, H. M. and Jacob, W. J. (1970). A regression method for obtaining real-time temperature and geopotential height profiles from satellite spectrometer measurements and its application to Nimbus 3 "SIRS" observations. *Mon. Weath. Rev.* **98**, 582–603.

Smith, W. L., Woolf, H. M., Hayden, C. M., Wark, D. Q. and McMillian, L. M. (1979). The TIROS-N operational vertical sounder. *Bull. Am. met. Soc.* **60**, 1177–1187.

2.4

Nowcasting Applications of Geostationary Satellite Atmospheric Sounding Data

W. L. SMITH, V. E. SUOMI, F. X. ZHOU and W. P. MENZEL

1 Introduction

On 9 September 1980 the first VAS was launched into geostationary orbit aboard the GOES-4 satellite. VAS is a second-order acronym standing for VISSR (visible and infrared spin scan radiometer) Atmospheric Sounder. VAS in orbit initiated a new era during which the space and time distribution of the temperature and moisture of the atmosphere will be observed with unprecedented detail. On 22 May 1981, a second VAS was orbited aboard the GOES 5.

The VAS is capable of achieving multi-spectral imagery of atmospheric temperature and water vapour evolution over short time intervals (~ 15 minutes) and achieving quantitative vertical sounding of the atmosphere with high spatial (~ 75 km) and temporal (~ 1 hour) resolution. Most important for nowcasting applications, the multi-spectral imagery and vertical sounding data can be received, processed, and made available to weather forecasters in real time.

The successful performance of VAS has already been demonstrated. Significant temporal and spatial variations of atmospheric temperature and moisture can be observed and retrieved with the anticipated accuracy of 1°C and 10%, respectively. As with other polar orbiting satellite (TOVS-TIROS Operational Vertical Sounder) soundings, the vertical resolution of the profiles is limited to several kilometres (Smith *et al.*, 1979). In this chapter we present a case study demonstrating the utility of this new geostationary satellite sounding capability for nowcasting, especially for the very-short-range prediction of intense local weather.

2 The VISSR Atmospheric Sounder (VAS)

The VISSR Atmospheric Sounder carried on the GOES is the result of a proposal by Suomi *et al.* (1971) to sound the atmosphere from a geostationary spacecraft so as to observe the variations of atmospheric water vapour and temperature distribution. It senses thermal radiation emission from atmospheric water vapour and carbon dioxide at various atmospheric depths as well as from the surface of the earth and clouds. The VAS is a radiometer possessing eight visible channel detectors and six thermal detectors that sense infrared radiation in 12 spectral bands (seven bands for sensing the temperature profile from the ground to the 50 mb level, three bands for sensing the water vapour concentration of the low, middle, and upper troposphere, and two bands for sensing the temperature of the surface of the earth and clouds). A filter wheel in front of the detector package is used to achieve the spectral selection. The central wavelengths of the spectral bands lie between 3.9 and 15 μm. Housed in the GOES satellite, VAS spins in a west to east direction at 100 r.p.m. and achieves spatial coverage at resolutions of 1 km in the visible and 7 or 14 km in the infrared (depending upon the detector employed) by stepping a scan mirror in a north-to-south direction (or vice versa).

Designed for multipurpose applications, the VAS can be operated in two different modes: (1) a Multi-Spectral Imaging (MSI) mode, and (2) a Dwell Sounding (DS) mode. Within each mode of operation there is a wide range of options regarding spatial resolution, spectral channels, spatial coverage, and the time frequency of observation. The mode of operation is programmed into an onboard processor from the ground through 39 processor parameters.

The DS mode of operation permits multiple samples of the upwelling radiance from a given earth swath in a given spectral band to be sensed by leaving the filter position and mirror position fixed during multiple spins of the spacecraft. The DS mode of operation was designed to achieve the improved signal-to-noise ratios required to interpret the spectral radiance measurements in terms of vertical temperature and moisture structure. Spatial averaging of several 14 km resolution observations may be employed to improve further the sounding radiance signal-to-noise ratio. Details of the VAS inflight performance are provided by Menzel (1981).

The MSI mode of operation is intended to achieve relatively frequent (e.g. half-hourly) full earth disc imagery of the atmospheric water vapour, temperature, and cloud distribution as well as variations in the surface skin temperature of the earth. In order to achieve full disc coverage at half-hourly intervals two modes of operation are possible: (a) four spectral channels can be observed (the visible at 1 km resolution, the 11 μm window at 7 km resolution, and two others at 14 km resolution) or (b) five spectral channels can be observed (the visible at 1 km resolution and any four infrared spectral channels at 14 km resolution).

For nowcasting applications a third mode of operation of the VAS has been programmed; the Dwell Imaging (DI) mode. The DI mode enables measurements in the tropospheric temperature and moisture sounding channels to be achieved over the North American region (20 to 55°N) every half hour. As will be shown later, the spectral coverage and accuracy of the data achieved in the DI mode is suitable for obtaining accurate quantitative measurements of lower and upper tropospheric relative humidity and geopotential thickness as well as atmospheric thermodynamic stability.

The VAS is capable of vertically sounding the atmosphere from a geostationary altitude of 36 000 km with about the same accuracy as that achieved by infrared sounders on polar orbiting spacecraft at an altitude of 1000 km. Initial results of the VAS vertical sounding capability have already been presented by Smith *et al.* (1981) and Smith and Woolf (1981). The novel capability of the VAS is its ability to sense the temporal variations in atmospheric temperature and moisture as well as the small-scale horizontal features.

Algorithms for deriving relative humidity and geopotential thickness have been developed (Smith and Zhou, 1981) to operate in real time on the University of Wisconsin McIDAS (Man–computer Interactive Data Access System) which ingests VAS data in real time. Also calculated in real time is the atmospheric stability defined in terms of the Total-Totals index: $TT = 2(T_{85} - T_{50}) - D_{85}$, where $(T_{85} - T_{50})$ is the difference between the temperatures at the 850 mb and 500 mb levels and D_{85} is the 850 mb level dewpoint depression. From VAS, the Total-Totals index is estimated from the lower tropospheric relative humidity and the 850–500 and 500–200 mb layer thicknesses calculated from the radiance observations. In general, each parameter is derived from an ensemble of twenty-five contiguous spatial samples (a 5×5 spatial matrix) of data providing a linear resolution of 75 km. To ensure freedom from cloud contamination, objective cloud checking and editing of the data is performed in the automated processing prior to the spatial averaging and calculation of meteorological parameters. After completion of the automated processing, a meteorologist at a McIDAS terminal rapidly inspects the results and subjectively eliminates any remaining erroneous data through cursor selection. For forecast applications, contour analyses of the results are then displayed over the imagery of preselected VAS channels.

3 A case study showing the mesoscale measurement capabilities of VAS

VAS data obtained on 20 July 1981 demonstrate the VAS nowcasting capabilities. Although half-hourly results were achieved, they are too numerous to present here; instead, three-hourly results are presented. The overall synoptic situation is shown in Fig. 1; full disk (MSI) 11 μm window and 6.7 μm H_2O images were obtained on 20 July at midday (1730 GMT). The 11-μm image shows that the United States is largely free of clouds except near the United States–Canadian border where a cold front persists. However, in the 6.7 μm upper tropospheric moisture image, a narrow band of moist air (delineated by

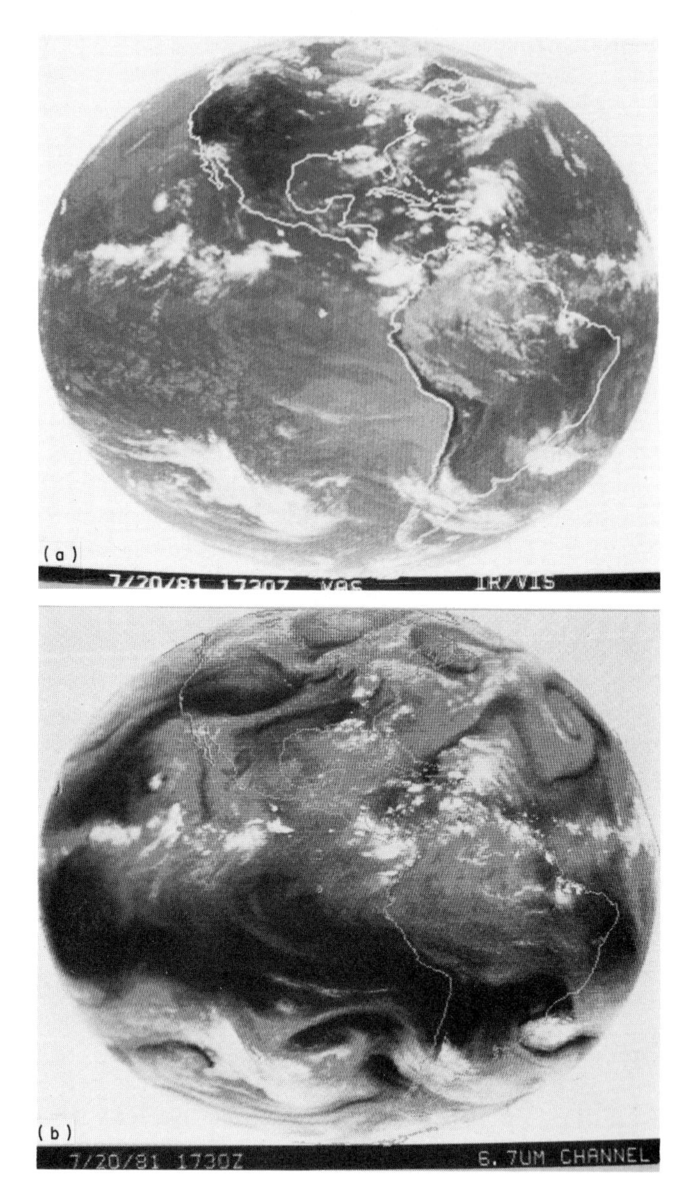

Fig. 1. Full disk images obtained on 20 July 1981 between 1730 and 1800 GMT: (a) 11-μm window and (b) 6.7-μm H_2O.

the low radiance greyish white areas of the image) stretches from the Great Lakes into the southwestern United States.

Figure 2 shows contours of derived upper tropospheric relative humidity superimposed over the VAS 6.7-μm brightness temperature images. In this case the narrow band of moist upper tropospheric air stretching across northern Missouri is the southern boundary of an upper tropospheric jet core (Fig. 6). The southeastward propagation of this moist band and associated jet core is seen. The bright cloud seen along the Illinois – Missouri border in the 21-GMT water vapour image corresponds to a very intense convective storm which developed between 18 and 21 GMT and was responsible for severe hail, thunderstorms, and several tornadoes in the St Louis, Missouri, region.

An objective of the VAS real-time parameter extraction software is to provide an early delineation of atmospheric stability conditions antecedent to intense convective storm development. For this purpose a Total-Totals stability index is estimated as described previously. On this day, the atmosphere was moderately unstable over the entire midwestern United States yet intense localized convection was not observed during the morning hours. In order to delineate regions of expected intense afternoon convection, three-hourly variations of stability (Total-Totals) were computed and displayed over the current infrared window cloud imagery on an hourly basis. Figure 3(a) shows the result for 18 GMT. A notable feature is the narrow zone of decreasing stability (positive three-hour tendency of Total-Totals) stretching from Oklahoma across Missouri and southern Illinois into western Indiana (the + 2 contour is drawn boldly in Fig. 3(a)). Also shown on Fig. 3(a) are the surface reports of thunderstorms which occurred between 21 and 23 GMT. Good correspondence is seen between the past three-hour tendency toward instability and the thunderstorm activity 3 to 5 hours ahead.

Figure 3(b) shows the one hour change in the Total-Totals index between 17 GMT and 18 GMT. The greatest 1-h decrease of atmospheric stability is along the border between northern Missouri and central Illinois where the tornado-producing storm developed during the subsequent three-hour period (see Fig. 2(d)). The stability variation shown was the largest one-hour variation over the entire period studied, 12–21 GMT.

A few detailed sounding results will now be presented for the Missouri region. The soundings were retrieved from VAS DS data with high spatial resolution (~ 75 km) using the interactive processing algorithms described by Smith and Woolf (1981). In Fig. 4 radiosonde observations of 700-mb temperature, and 700-mb temperatures retrieved from GOES-5 VAS DS radiance data, are presented for a region surrounding the state of Missouri on 20 July 1981. Figure 5 shows a similar comparison for the 300-mb dewpoint temperature. Even though the VAS values are actually vertical mean values for layers centred about the indicated pressure levels, it can be seen from the 12 GMT observations that the VAS is broadly consistent with the available radiosondes while at the same time delineating important small-scale features which cannot be resolved

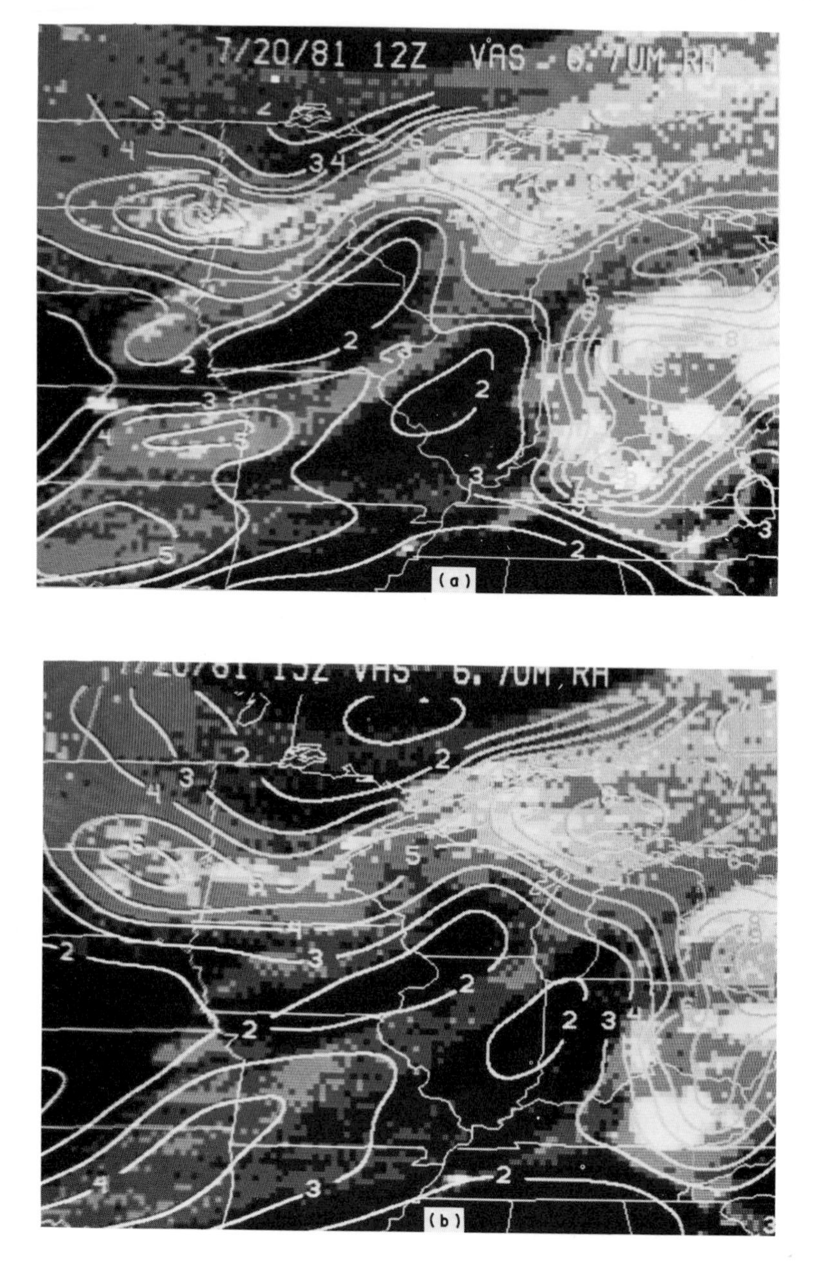
7/20/81 12Z VAS 6.7UM RH
(a)
VAS 6.7UM RH
(b)

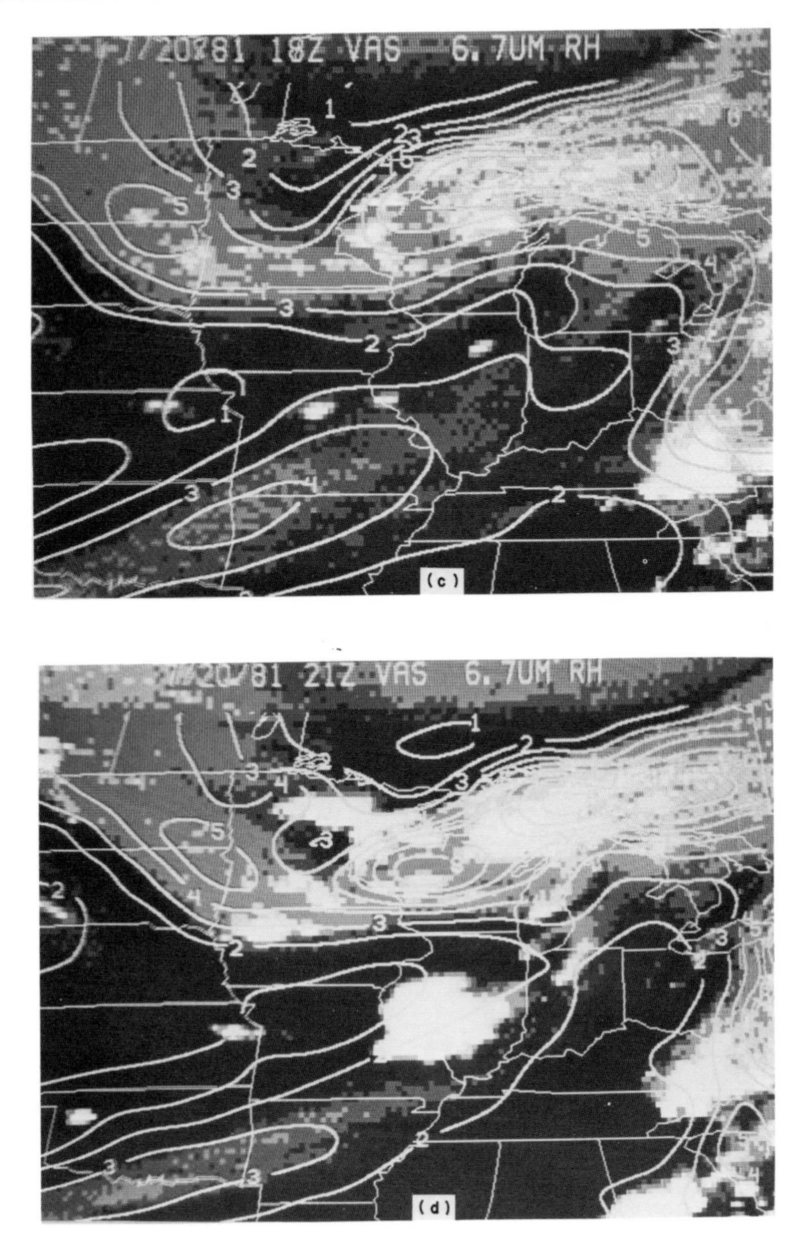

Fig. 2. Upper tropospheric relative humidity (%/10) over the VAS image of 6.7-μm atmospheric water vapour radiance emission: (a) 12 GMT; (b) 15 GMT; (c) 18 GMT; and (d) 21 GMT on 20 July, 1981.

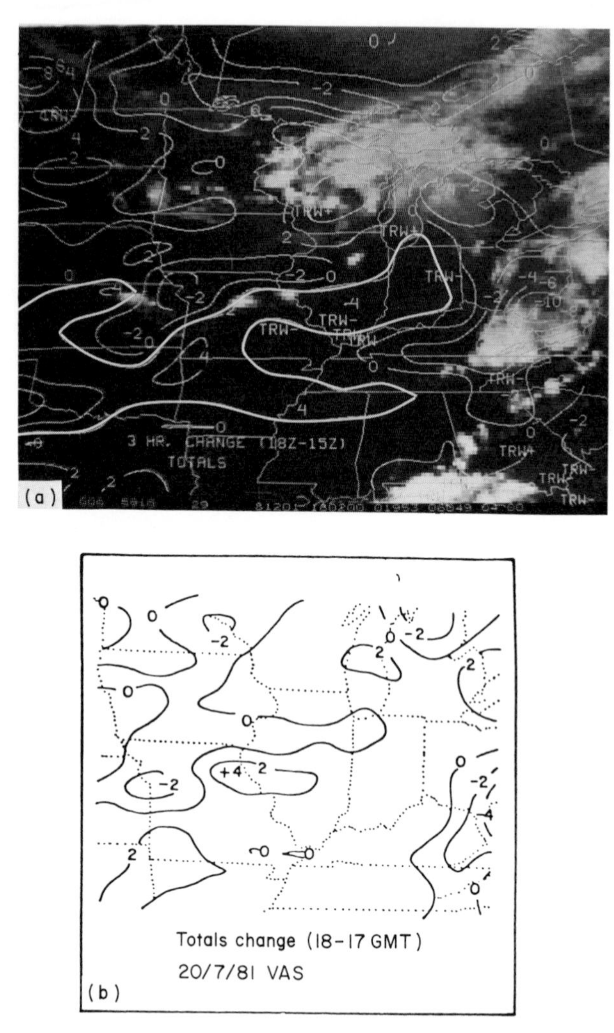

Fig. 3. (a) Three-hour variation of VAS derived Total-Totals index (°C) between 15–18 GMT superimposed over the 18-GMT VAS 11-μm image of cloudiness. The symbols of thunderstorms (TRW) which were observed between 20 and 23 GMT are also shown, (b) 1-hour (17–18 GMT) variation of Total-Totals index (°C) showing that the maximum one-hour variation (4°C) observed by VAS on 20 July 1981 occurred at the location of and just prior to the development of a severe convective storm.

by the widely spaced radiosondes. This is particularly obvious in the case of moisture (Fig. 5) where the horizontal gradients of dewpoint temperature are as great as 10°C over a distance of less than 100 km.

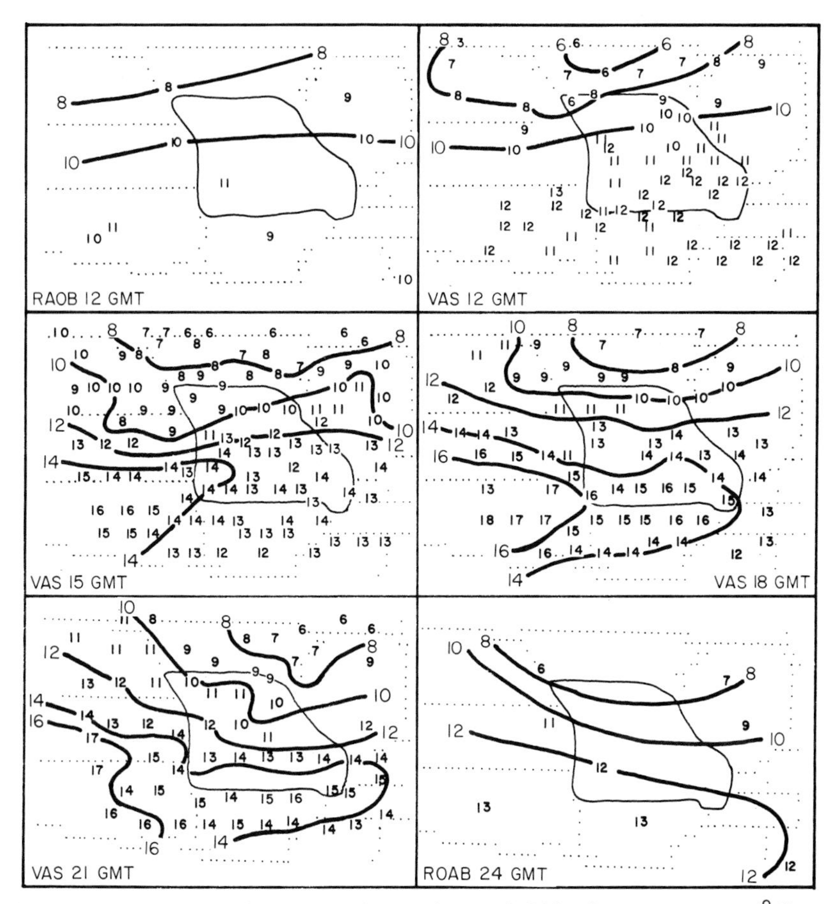

Fig. 4. Radiosonde and VAS observations of 700 mb temperature (°C) on 20 July, 1981.

The temporal variations of the atmospheric temperature and moisture over the Missouri region are observed in detail by the three-hour interval VAS observations presented in Figs. 4 and 5. For example, in the case of the 700-mb temperature, the VAS observes a warming of the lower troposphere with a tongue of warm air protruding in time from Oklahoma across southern Missouri and northern Arkansas, the maximum temperatures being observed around 18 GMT in this region. Although we believe this to be a real diurnal effect, undected by the 12-h interval radiosonde data, it is possible that it has been exaggerated by the influence of high skin surface temperature. This problem needs to be investigated further. The temporal variation of upper tropospheric moisture may be seen in Fig. 5; a narrow intense moist tongue (high dewpoint temperature) across northwestern Missouri steadily propagates southeastward

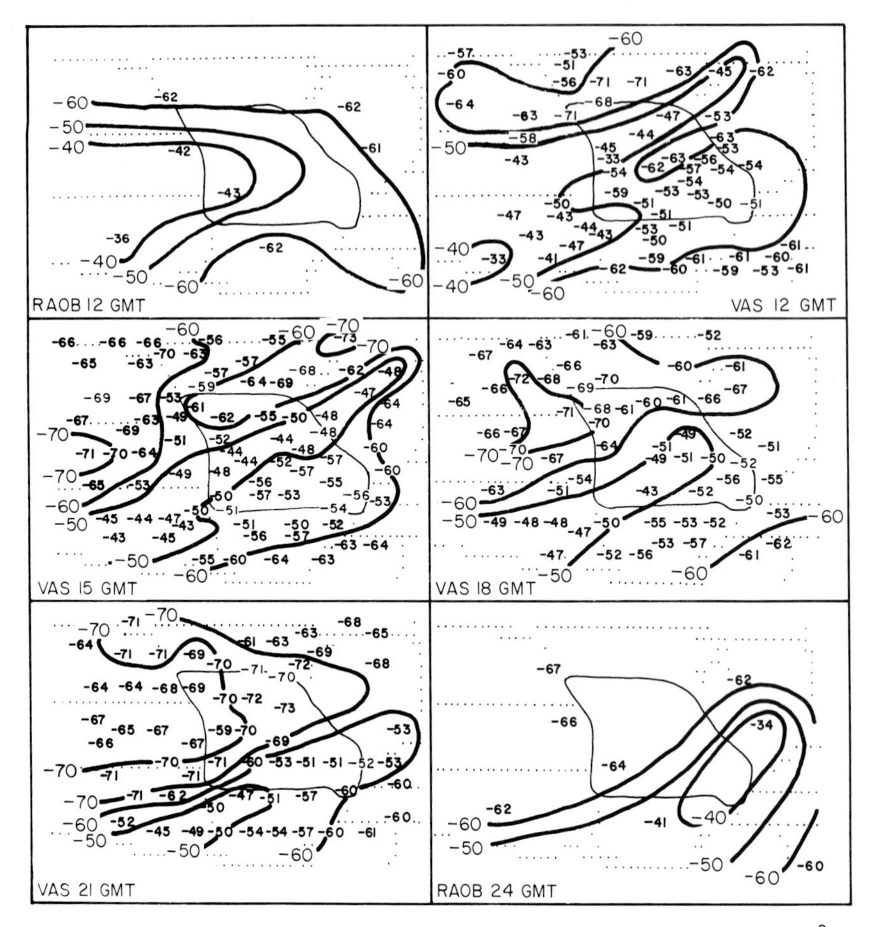

Fig. 5. Radiosonde and VAS observations of 300-mb dewpoint temperature (°C) on 20 July 1981.

with time. The dewpoint temperature profile results achieved with the interactive profile retrieval algorithm are consistent with the layer relative humidities achieved automatically in real time (Fig. 2). It is noteworthy that over central Missouri, where the intense convective storm developed, the VAS observed a very sharp horizontal gradient of upper tropospheric moisture just prior to and during the storm genesis between 18 and 21 GMT. Here again the inadequacy of the radiosonde network for delineating important spatial and temporal features is obvious. The discrepancy between the 21 GMT VAS and 24 GMT radiosonde observation over southern Illinois (Fig. 5) is due to the existence of deep convective clouds. A radiosonde observed a saturated dewpoint value of − 34°C in the cloud while VAS was incapable of sounding through the heavily

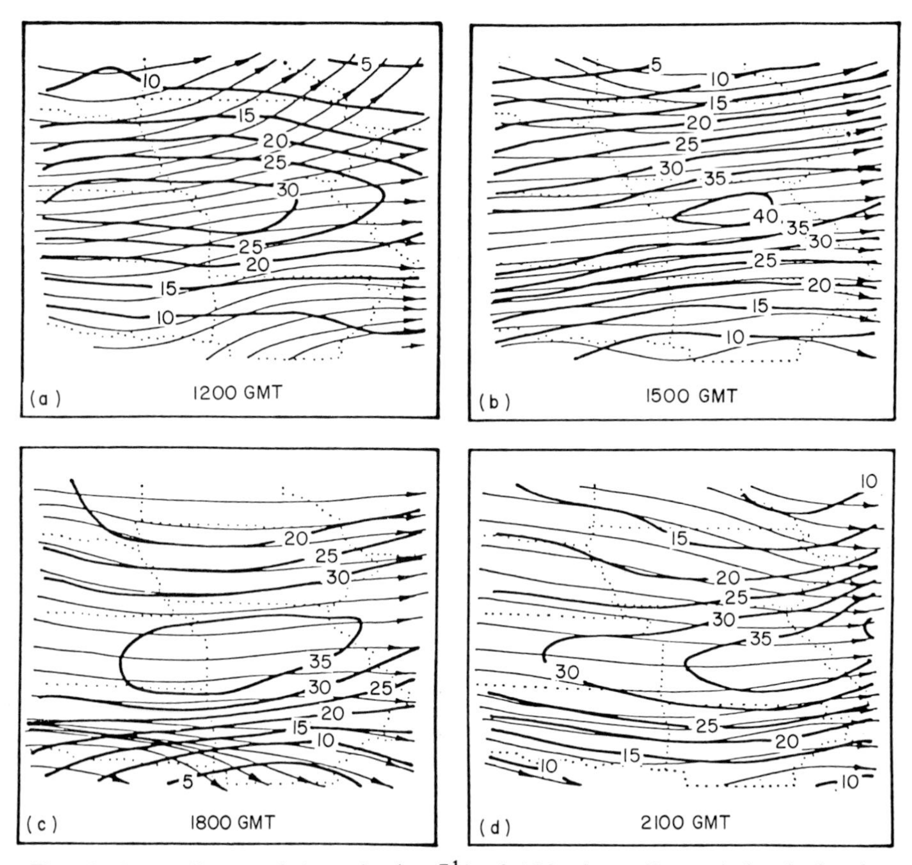

Fig. 6. Streamlines and isotachs ($m\,s^{-1}$) of 300 mb gradient wind calculated from VAS temperature soundings: (a) 12 GMT, (b) 15 GMT, (c) 18 GMT, and (d) 21 GMT on 20 July, 1981.

clouded area. Nevertheless we believe that the VAS DS data provide, for the first time, the kind of atmospheric temperature and moisture observations needed for the timely initialization of a mesoscale numerical model for predicting localized weather.

Figure 6 shows streamlines and isotachs of 300-mb gradient winds derived from the VAS temperature profile data, where the curvature term was approximated from the geopotential contours. There is agreement (not shown) between the VAS 12 GMT gradient winds and the few radiosonde observations in this region. As can be seen there is a moderately intense subtropical jet streak which propagates east-southeastward with time. Note that at 18 GMT the exit region of the jet is over the area where the severe convective storm developed. As shown by Uccellini and Kocin (1981), the mass adjustments and isallobaric forcing of a

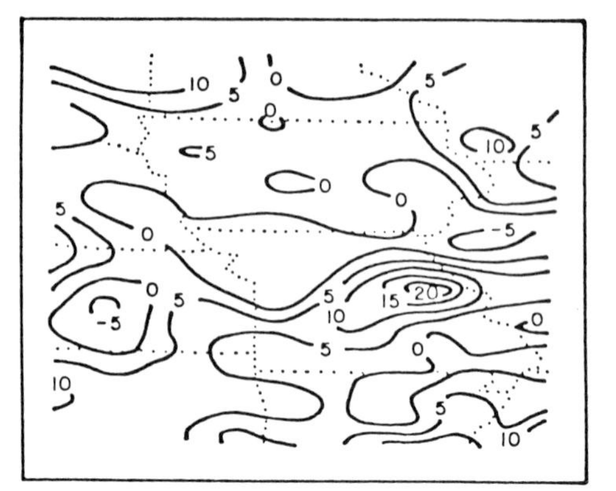

Fig. 7. Three hour change of precipitable water vapour (mm) between 15 and 18 GMT on 20 July, 1981.

low level jet produced under the exit region of an upper tropospheric jet streak can lead to rapid development of a convectively unstable air mass within a 3- to 6-h time period.

Figure 7 shows the three-hour change of total precipitable water prior to the severe convective storm development. The instantaneous fields from which Fig. 7 was derived showed a number of short-term variations. Nevertheless it is probably significant that a local maximum exists over the location of the St Louis storm prior to its development. It is suspected that the convergence of lower tropospheric water vapour is a major mechanism for the thermodynamic destabilization of the atmosphere leading to severe convection between 18 and 21 GMT along the Missouri–Illinois border.

4 Conclusion

The VAS geostationary satellite sounder offers exciting new opportunities for real-time monitoring of atmospheric processes and for providing on a timely basis the vertical sounding data at the spatial resolution required for initializing mesoscale weather prediction models. Results from this case study and others not reported here suggest that VAS can detect, several hours in advance, the temperature, moisture, and jet streak conditions forcing severe convective development. These preliminary results indicate broad agreement with conventional data; however, there are areas of uncertainty and more detailed analysis will be required to clarify them.

Unfortunately, in order to achieve the Dwell Imaging or Dwell Sounding multi-spectral data, the VAS cannot at the same time achieve full disk cloud imagery with half-hour frequency as has been the past operational practice.

Consequently, a new geostationary satellite operating scenario is required. A program is being initiated by NESS to steadily integrate new VAS capabilities into daily operational practice. This will provide meteorologists with much more information about the varying state of the atmosphere than is provided by half-hourly cloud imagery.

Acknowledgements

Thanks are extended to G. S. Wade and A. J. Schreiner for their continual support in processing the VAS data. Appreciation for software assistance is extended to H. M. Woolf and C. M. Hayden of NOAA/NESS. Susan Francis and Vong Pui-sui assisted in the preparation of the manuscript.

References

Menzel, W. P. (1981). Postlaunch study report of VAS-D performance. Report to NASA under contract NAS5-21965 from SSEC, Space Science and Engineering Center, Madison, Wisconsin.

Smith, W. L. (1970). Iterative solution to the radiative transfer equation for the temperature and absorbing gas profile of an atmosphere. *Appl. Opt.* **9**, 1993–1999.

Smith, W. L., Woolf, H. M., Hayden, C. M., Wark, D. Q., and McMillin, L. M. (1979). The TIROS-N Operational Vertical Sounder. *Bull. Am. met. Soc.* **60**, 1177–1187.

Smith, W. L., Suomi, V. E., Menzel, W. P., Woolf, H. M., Stromovsky, L. A., Revercomb, H. E., Hayden, C. M., Erickson, D. N. and Mosher, F. R. (1981). First sounding results from VAS-D. *Bull. Am. met. Soc.* **62**, 232–236.

Smith, W. L. and Woolf, H. M. (1981). Algorithms used to retrieve surface-skin temperature and vertical temperature and moisture profiles from VAS radiance observations. Submitted to *Mon. Weath. Rev.* for publication.

Smith, W. L. and Zhou, F. X. (1981). Rapid extraction of layer relative humidity, geopotential thickness, and atmospheric stability from satellite sounding radiometer data. Submitted to *Appl. Opt.* for publication.

Suomi, V. E., VonderHaar, T., Krauss, R. and Stamm, A. (1971). Possibilities for sounding the atmosphere from a geosynchronous spacecraft. *Space Res.* **XI**, 609–617.

Uccellini, L. W. and Kocin, P. J. (1981). Mesoscale aspects of jet streak coupling and implications for short term forecasting of severe convective storms. *In* "Nowcasting: Mesoscale observations and short-range prediction", pp. 375–380. Proceedings of a symposium at the IAMAP General Assembly, 25–28 August 1981, Hamburg. European Space Agency, ESA SP-165.

2.5

Nowcasting the Position and Intensity of Jet Streams Using a Satellite-borne Total Ozone Mapping Spectrometer

M. A. SHAPIRO, A. J. KRUEGER and P. J. KENNEDY

1 Introduction

The temporal evolution of the tropopause during extratropical cyclogenesis and the formation of upper-tropospheric jet streams and frontal zones has been studied by Reed (1955), Danielsen (1968), Shapiro (1978, 1980), and Shapiro *et al.* (1980). These investigations established the use of potential vorticity and ozone for locating the tropopause. With measurements from research aircraft, it was shown that both potential vorticity and ozone within the lower stratosphere are horizontally and vertically displaced by actions of three-dimensional air motions near the tropopause. The vertical deformation of the tropopause produces large horizontal gradients in tropopause height, potential vorticity and ozone in the stratospheric layer beneath the 100-mb pressure level. However, because of the limited altitude range of research aircraft, it has not been possible until recently to ascertain the extent to which these horizontal gradients of ozone below 100 mb would modulate the total columnar amount of ozone and thus give rise to a significant signature in total atmospheric ozone resulting from large gradients in tropopause height across jet streams and extra-tropical cyclones.

The first sensor capable of observing the modulation of total ozone by deformations of the tropopause is the Total Ozone Mapping Spectrometer (TOMS), a research instrument flown on the Nimbus 7 satellite. In the present chapter, we present case studies which compare TOMS observations with conventional observations and research aircraft measurements to illustrate the correlations between mesoscale structure in TOMS ozone maps and the geographical location of jet-stream systems near the tropopause. In addition, we discuss the nowcasting potential of the TOMS observing system.

2 Instrumentation

The Nimbus 7 satellite was launched in a local noon–midnight sun-synchronous polar orbit on 24 October 1978. The TOMS instrument is an Ebert–Fastie monochromator with six, serially sampled, fixed exit slits at wavelengths between 312.5 and 280.0 nm (Heath *et al.*, 1975). The atmospheric albedo in this wavelength span is determined by a varying blend of ozone absorption, Rayleigh scattering and surface (and cloud) reflections. By judicious selection of wavelengths, these effects are separable, and total ozone can be determined with a precision of 2% relative to ground-truth stations. To map the ozone, the $3^\circ \times 3^\circ$ instrument field of view is swept across the track so that contiguous sampling is obtained between scans and between orbits (see Nimbus Project, 1973) This yields a spatial resolution of 50×50 km at the nadir and 75×150 km at the edges of the swath. The data processing algorithms were developed and the data were validated by the Solar Backscatter Ultraviolet/TOMS Nimbus Experimental Team and the Ozone Processing Team at Goddard Space Flight Center. False colour global images of the total ozone data as shown in Frontispiece (1) were developed by the Instrument Evaluation Branch at Goddard Space Flight Center.

Meteorological and *in situ* chemical measurements for this study were made with the NCAR Sabreliner research aircraft. The reader is referred to Newcomer *et al.* (1973) and Shapiro (1980) for further discussion of the instrumentation and research application of this aircraft.

3 Case study examples

We now present two examples which illustrate the usefulness of the TOMS data for delineating the geographical location and intensity of jet-stream systems near the tropopause. These examples are taken from two consecutive days during which meteorological and ozone measurements were made from the NCAR Sabreliner research aircraft.

The first example is from 0000 GMT 4 April 1981, when the Sabreliner was gathering jet-stream data over Texas and Oklahoma. Figure 1 presents the analysis of the 250-mb wind field which has incorporated data from rawinsondes, commercial aircraft pilot reports, satellite cloud motion winds, and the Sabreliner. The analysis shows a southwesterly jet stream which crosses the Baja Peninsula, enters west Texas and weakens as it passes over the Great Lakes. The maximum wind speeds exceed $75\ \mathrm{m\,s^{-1}}$ at the jet core. A second but weaker northerly jet system is seen entering the northwest United States coast over Washington and Oregon.

The TOMS ozone observations for the synoptic situation of Fig. 1 are shown in Fig. 2. The ozone analysis, which applies at local noon (6 hours before 0000 GMT over the central United States), reveals ozone isopleths which are oriented nearly parallel to the wind direction with the strongest gradient located in the vicinity of the core of the south-westerly jet stream current of Fig. 1. The ozone gradient is concentrated on the mesoscale (~ 100 km) and broadens and weakens as the wind speeds lessen in the direction up and down stream from the jet-stream wind maximum. The largest gradient in ozone is found in the region of strongest jet stream wind speeds. Larger-scale features of the ozone distribution are the high values associated with the trough axis over the west-central United States and the lower values within the anticyclonic flow to the east, which result from the low and high tropopause altitudes within these respective geographic regions.

On the following day the jet structure was quite similar to that observed at 0000 GMT 4 April 1981. The 250-mb wind velocity analysis for 0000 GMT 5 April 1981 (Fig. 3) shows the strongest portion of the jet stream over the central United States in the south-westerly flow ahead of a large-amplitude wave. On this day, the NCAR Sabreliner performed a series of horizontal traverses at levels between 540 and 180 mb along the line $\overline{AA'}$ of Fig. 3, in order to map the vertical distribution of wind speed, potential temperature and ozone across this jet-front system.

Figure 4 presents the wind speed, potential temperature and tropopause analysis of the observations taken in two Sabreliner flights over the time interval 1645 GMT 4 April 1981 and 0030 GMT 5 April 1981. Rawinsonde data at 0000 GMT 5 April 1981 were incorporated into this cross-section analysis. The tropopause is defined by the near zero-order discontinuity in potential vorticity which has been shown to separate air of stratospheric versus tropospheric origin (Reed, 1955; Shapiro, 1980).

The cross-section analysis shows the merging of a 75 m s^{-1} polar jet with a greater than 80 m s^{-1} subtropical jet between Oklahoma City (OKC) and Stephenville (SEP). A strong mid-tropospheric front extends downward from the stratosphere beneath the two jets. The potential vorticity tropopause folds deeply into the mid-tropospheric front with a secondary weaker folding to the cyclonic shear side of the subtropical jet.

The vertical distribution of ozone for the flight section of Fig. 3 was obtained from a DASIBI ozone analyser on board the Sabreliner. The results (Fig. 5) show the characteristic intrusion of ozone-rich air into the two frontal layers. The folds in the ozone (Fig. 5) are coincident with the multiple folding of the tropopause. An important result shown in Figs. 4 and 5 is horizontal discontinuity in both the depth of the stratosphere and ozone which is seen to occur across this jet system. The ozone analysis (Fig. 5) shows the high tropopause (190 mb) at Stephenville (SEP) with little ozone below it. In contrast, the tropopause over

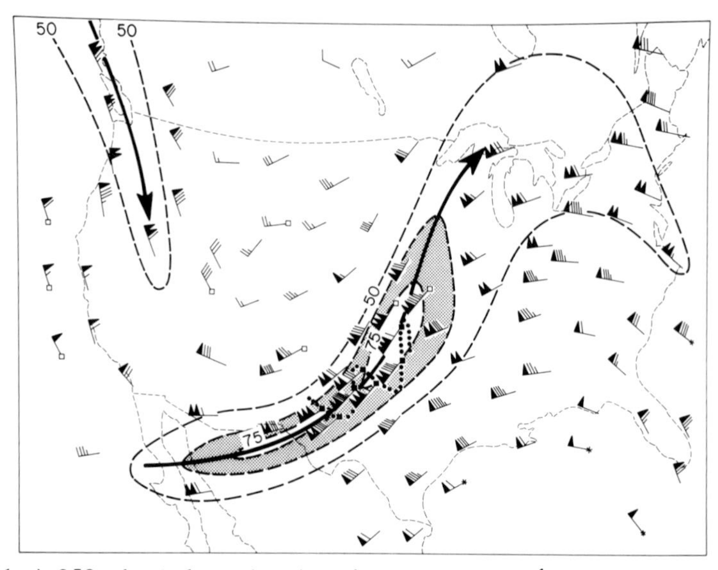

Fig. 1. A 250-mb wind speed analysis (dashed lines, m s^{-1}) at 0000 GMT 4 April 1981. Rawinsonde wind velocities (flag, 25 m s^{-1}; barb, 5 m s^{-1}; half barb, 2.5 m s^{-1}), commercial airline wind reports (open box vectors), satellite cloud motion winds (asterisked vectors), Sabreliner research aircraft winds (closed box vectors), Sabreliner flight track (heavy dots), jet stream axes (heavy solid arrows) and 62.5 to 75.0 m s^{-1} wind speed interval (stippled area).

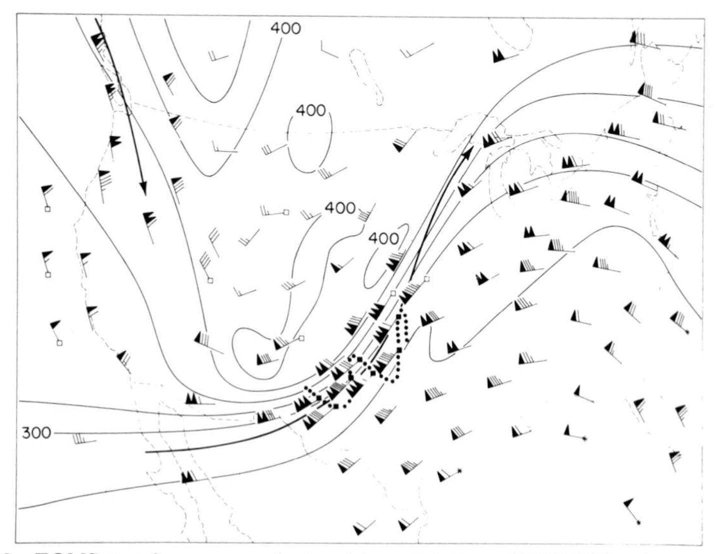

Fig. 2. TOMS total ozone analysis, thin solid lines (1000 DU = 1 atm cm), at local noon for the situation depicted in Fig. 1. Wind vectors and jet axes, exactly the same as Fig. 1.

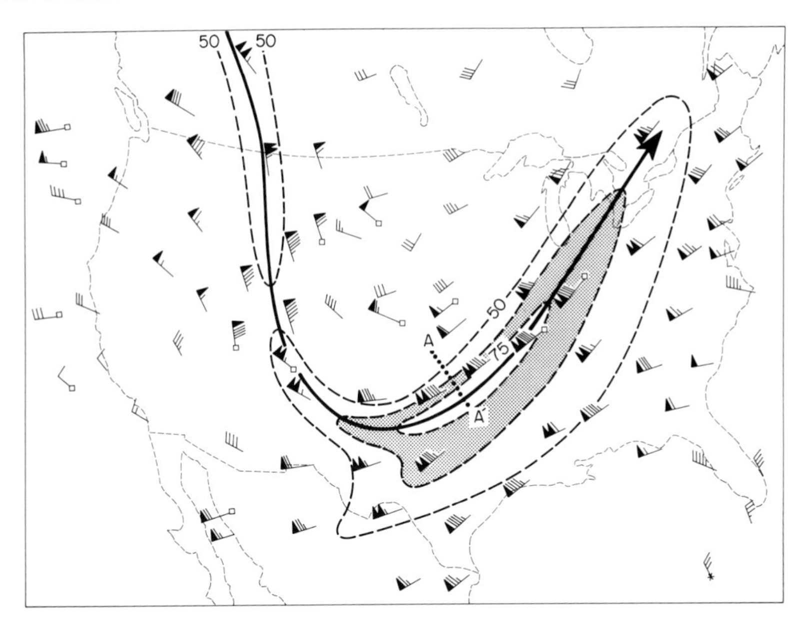

Fig. 3. A 250-mb wind speed analysis at 0000 GMT 5 April 1981. $\overline{AA'}$ projection line for Fig. 4 cross-section. Isopleths, wind vectors and jet axes, same format as Fig. 1.

Oklahoma City (OKC) is at 430 mb with a deep pool of ozone-rich air above it.

We next look at the TOMS ozone map for evidence of mesoscale discontinuities in total ozone which result from large differences in the depth of the stratosphere across the jet system depicted in Figs. 4 and 5. Figure 6 presents the TOMS ozone map for local noon (1800 GMT over the central United States) on 4 April 1981. This analysis is to be compared with the jet-stream analysis at 250 mb in Fig. 3. The comparison reveals that the large mesoscale gradient in total ozone over the central United States coincides with the axis of the 250-mb jet-stream current. The intensity of the gradient appears to be proportional to the magnitude of the jet wind speed. Current research on additional case studies will attempt to characterize this relationship.

4 Nowcasting applications of TOMS ozone observations

This chapter has presented evidence which suggests that wave structure and horizontal gradients in tropopause height give rise to modulations in total atmospheric ozone which are identifiable in ozone maps from TOMS. In the

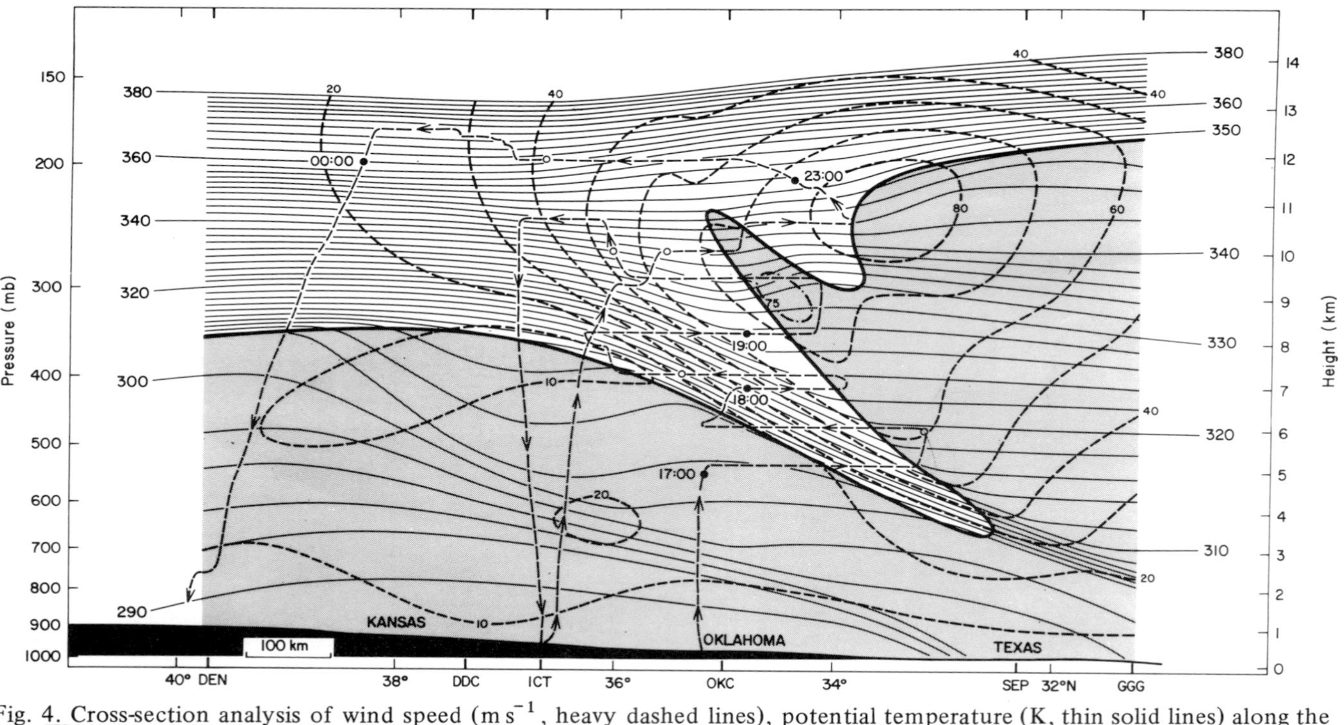

Fig. 4. Cross-section analysis of wind speed ($m\,s^{-1}$, heavy dashed lines), potential temperature (K, thin solid lines) along the line $\overline{AA'}$ for 1900 GMT 4 April 1981. Sabreliner flight track (thin dashed lines) with hourly (solid circles) and half hourly (open circles) time hacks. Tropopause ($10^{-5}\,K\,mb^{-1}\,s^{-1}$ isopleth of potential vorticity, heavy solid line). Troposphere, stippled area. Rawinsonde soundings are from Denver, Colorado (DEN), Dodge City, Kansas (DDC), Oklahoma City, Oklahoma (OKC), Stephenville, Texas (SEP) and Longview, Texas (GGG).

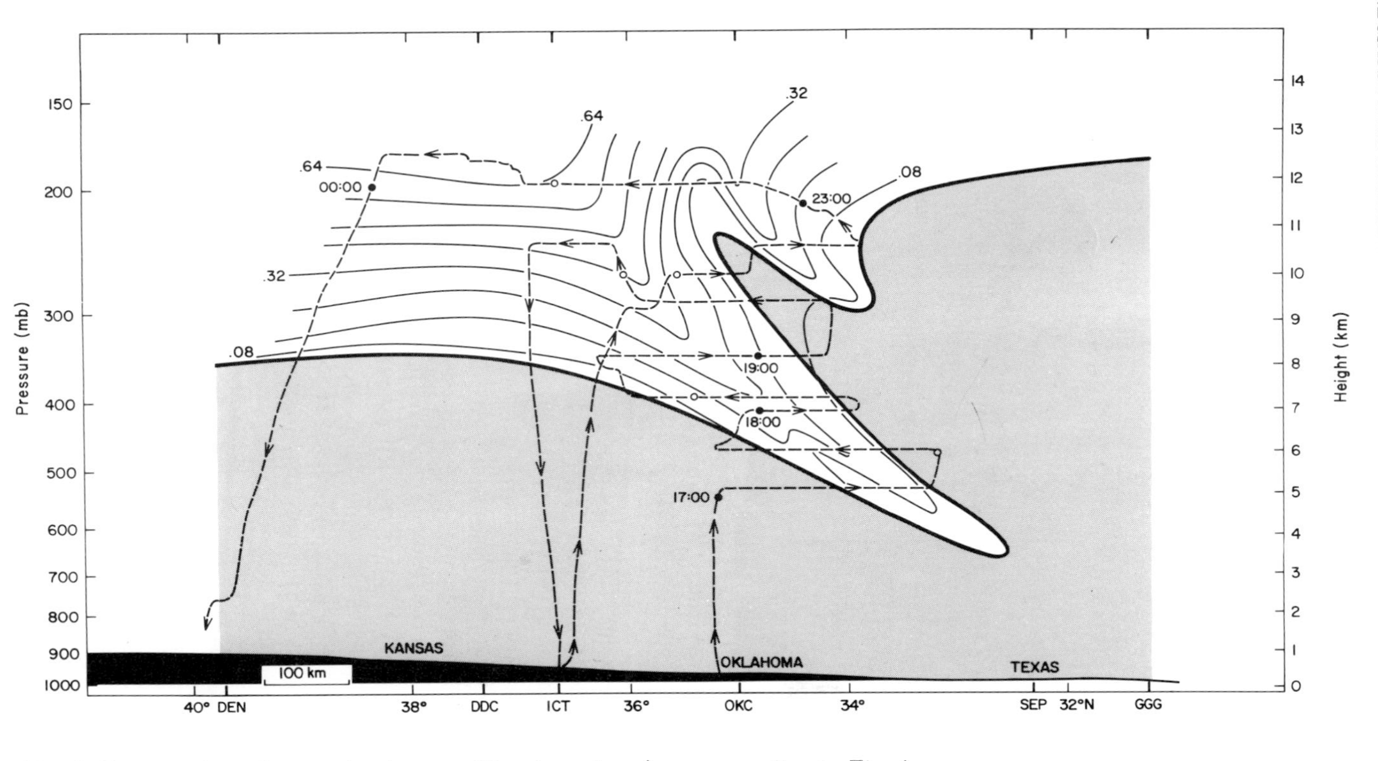

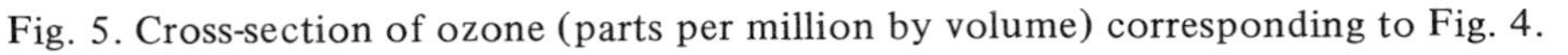
Fig. 5. Cross-section of ozone (parts per million by volume) corresponding to Fig. 4.

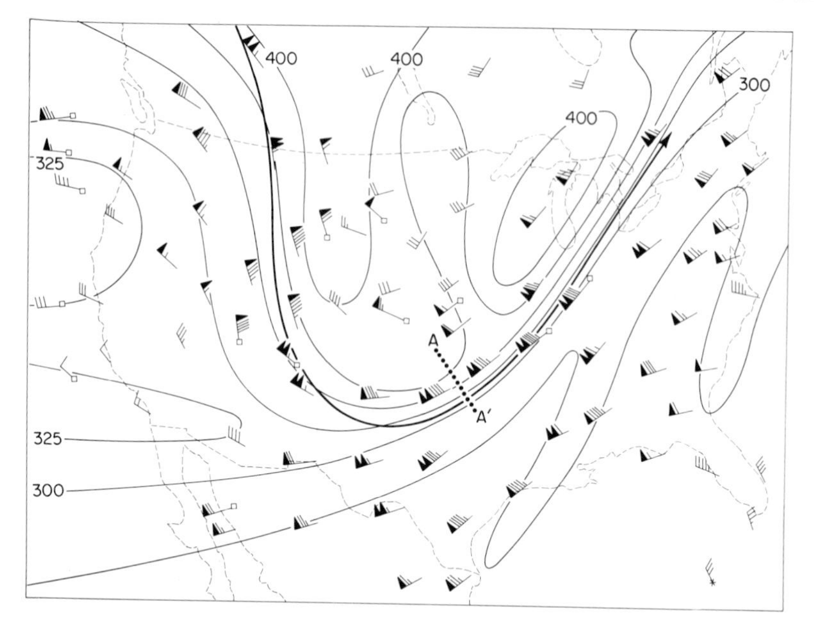

Fig. 6. TOMS total ozone for the situation depicted in Fig. 3. Isopleths, wind vectors and jet axis, same format as Fig. 2. The area covered here is the central portion in Frontispiece (1).

light of this evidence, it is proposed that TOMS data be applied to nowcasting the location and intensity of atmospheric flows in the vicinity of the tropopause. The TOMS observations have the potential for providing valuable information about jetstream characteristics, particularly over the data-sparse regions of the northern hemispheric oceans, and over the Southern Hemisphere. Because of the accelerating costs in aviation fuel, the authors foresee the application of TOMS to the problem of specifying minimum flight time routes of commercial aircraft.

It remains for future research to quantify the relationship between modulations in total atmospheric ozone and weather systems near the tropopause, such that this new data base may be incorporated into the operation loop of short-range (< 36-h) numerical weather prediction.

References

Danielsen, E. F. (1968). Stratospheric–tropospheric exchange based on radioactivity, ozone and potential vorticity. *J. atmos. Sci.* **25**, 502–518.

Heath, D. F., Krueger, A. J., Roder, H. A. and Henderson B. D. (1975). The solar back scatter ultraviolet and total ozone mapping spectrometer (SBUV/TOMS) for Nimbus G. *Opp. Engr.* **14**, 323–331.

Newcomer, L. E. and Ruth, R. (1973). The NCAR Sabreliner. Atmospheric Technology 1, pp. 28–30. National Center for Atmospheric Research, Boulder, Colorado.

Nimbus Project (1978). The Nimbus 7 users guide. Landsat Nimbus 7 Project Office, Goddard Space Flight Center, Greenbelt, Maryland 20771, USA.

Reed, R. J. (1955). A study of a characteristic type of upper-level frontogenesis. *J. Met.* **12**, 226–237.

Shapiro, M. A. (1978). Further evidence of the mesoscale and turbulent structure of upper-level jet stream-frontal zone systems. *Mon. Weath. Rev.* **106**, 1100–1111.

Shapiro, M. A. (1980). Turbulent mixing within tropopause folds as a mechanism for the exchange of chemical constituents between the stratosphere and troposphere. *J. atmos. Sci.* **37**, 994–1004.

Shapiro, M. A., Reiter, E. R., Cadle, R.D., and Sedlacek, W. A. (1980). Vertical mass- and trace constituent transports in the vicinity of jet streams. *Arch. Met. Geophys. Bioklim., Ser. B,* 193–206.

PART 3

Simple Forecasting Methods

Introduction

The various forms of mesoscale observational data described earlier in this book can be used by a forecaster in two quite distinct ways. One approach, discussed later in Part 4, is to use the detailed observations as input to numerical–dynamical models. That approach is attractive as a long-term objective, but there are practical difficulties that will delay its widespread implementation. In the meantime much can be gained from a conceptually simple approach involving the detailed description of the current weather pattern and extrapolating it for a few hours ahead. This approach is not as straightforward as it might seem, however, because the remote probing techniques that underpin this approach do not provide exactly the information that the forecaster needs. Therefore one of the first tasks is to interpret the data to give a meaningful analysis. This is the topic of Chapters 3.1 and 3.2.

Purdom (Chapter 3.1) explains how the subjective interpretation of cloud imagery from a geostationary satellite can play a central role in nowcasting by enabling the forecaster to diagnose the dynamic and thermodynamic processes taking place on the mesoscale. Such is the complexity of mesoscale interactions, especially in convective situations, that the resulting increase in forecast skill is liable to fall short of the expectations the reader might derive from case studies viewed after the event. Nevertheless there is good reason to expect significant improvement in very-short-range forecasts to stem from the deeper understanding of the current weather that the forecaster can gain from animated sequences of up-to-date cloud and radar images.

Satellite imagery is often available from two or more channels simultaneously. Liljas (Chapter 3.2) describes how cloud types and probable precipitation intensity, as well as land/sea boundaries, can be assessed better by means of a combined analysis of the data from three channels of a TIROS-N type polar orbiting satellite. This is difficult to do subjectively and in any case the forecaster would be inundated with data. Accordingly an objective procedure is outlined by means of which the multispectral information can be automatically distilled into a single image which corresponds to a useful cloud and precipitation classification.

Whereas Purdom and Liljas deal with the interpretation of current observations, Austin and Bellon (Chapter 3.3) are concerned with very-short-range forecasts obtained by extrapolating current observational data after they have been analysed in terms of the wanted parameter. The parameter considered in this case is precipitation as derived from ground-based radar or from a bispectral analysis of cloud imagery from a geostationary satellite. The approach, which is being used operationally, is objective and fully automated.

Forecasts based upon simple linear extrapolation become increasingly unreliable as the lead time extends beyond a few hours because of the neglect of development and decay. Terrain-induced developments can be assessed with the aid of mesoscale numerical models as discussed in Part 4. Developments associated with the natural life-cycle of the mesoscale events are more difficult to handle using numerical models, but a forecaster can improve his understanding of the situation and hence his forecast by the subjective application of conceptual models of the structure and evolution of mesoscale weather systems. This approach is discussed by Zipser (Chapter 3.4), who illustrates this principle by considering the life-cycle of mesoscale convective systems of a kind found in both the Tropics and mid-latitude regions.

K. A. Browning

3.1 Subjective Interpretation of Geostationary Satellite Data for Nowcasting

JAMES F. W. PURDOM

1 Introduction

Nowcasting and the very-short-range forecasting of mesoscale weather is one of the most challenging problems in meteorology today. There are a variety of reasons for this, not the least of which is that the mesoscale remains ill-defined and poorly understood. For that reason, immediate improvements are most likely to be achieved through the incorporation of empirical information into new procedures for nowcasting various meteorological situations. This is no trivial task. Such an approach requires immediate display capability for current data (satellite, radar, surface mesonets, etc.), provision for their combination, and the availability of products derived from those data in a readily comprehensible form for the meteorologist involved in the decision-making process. This is one of the major goals of the PROFS program (elsewhere in this volume, or Beran, 1980). The broader problem is the investigation of how satellite and other data sets can be combined and analysed to gain a better understanding of mesoscale atmospheric processes. Satellite data and their use for nowcasting and very-short-term forecasting, with particular emphasis on convection, is the major topic to be discussed in this chapter.

2 Geostationary satellite data as the key to nowcasting and very-short-range forecasting

2.1 A BRIEF DESCRIPTION OF GEOSTATIONARY SATELLITES

Currently there are five geostationary meteorological satellites in orbit around the globe, all of which provide routine meteorological observations. GOES (Geostationary Operational Environmental Satellite) is the current operational geostationary meteorological satellite in use by the United States. Two GOES are operated routinely: one over the East Coast of the United States, and the other between Hawaii and the West Coast of the United States. The heart of the GOES satellite is its Visible and Infrared Spin Scan Radiometer (VISSR) which senses the equivalent black body temperature of the scene beneath it with a spatial resolution of 8 km by day and night, as well as 1 km resolution visible images during the day, both on a nominal half-hourly basis (for further information on GOES, see Ludwig, 1974). The information given in the following sections of this chapter is based on GOES imagery. Recently the next generation geostationary satellite, GOES/VAS, has been launched. GOES/VAS is similar to the present GOES with one important exception – its VISSR can also act as an Atmospheric Sounder; thus the acronym GOES/VAS. The sounder portion of GOES/VAS is in the demonstration phase, and preliminary results look promising (see W. L. Smith *et al.*, this volume, pp. 123–135, or Smith *et al,* 1981).

Japan, India and Europe also have operational geostationary satellites that provide routine meteorological observations. The European satellite, called METEOSAT, is located near the 0° meridian. The sensor onboard METEOSAT observes the earth, its cloud cover and upper tropospheric water vapour using three spectral bands. They are: (1) visible to near infrared (0.4 to 1.1 μm) at a resolution of 2.5 km; (2) mid to upper level water vapour (5.7 to 7.1 μm) at a resolution of 5 km; and the infrared window (10.5 to 12.5 μm) with a resolution of 5 km. Data are taken of the full earth disk once every 30 minutes. The Japanese satellite, called GMS for Geostationary Meteorological Satellite, is located near 130° East. The satellite is similar in characteristics to the GOES satellites. The observational frequency of GMS varies depending on the purpose for which the observations are being taken. Further information on METEOSAT and GMS may be found in Mosher (1980). The Indian meteorological satellite, INSAT, was launched in April 1982. The satellite is located near 80° E and has an imaging capability similar to GOES.

2.2 SATELLITE DATA, THE KEY TO NOWCASTING

The geostationary satellite has the *unique ability* to frequently observe the atmosphere (sounders) and its cloud over (visible and infrared) from the synoptic scale down to the cloud scale. This ability to provide frequent, uniformly calibrated data sets from a single sensor over a broad range of meteorological scales places the geostationary satellite at the very heart of the understanding, nowcasting and very-short-range forecasting of mesoscale weather development. The clouds and cloud patterns observed in a satellite image or animated series of images represents the integrated effect of ongoing dynamic and thermodynamic processes in the atmosphere. When that information is combined with radar, surface and upper air observations, then many of the important processes in mesoscale weather development may be better analysed and understood. It is from this better analysis and understanding of mesoscale processes that improved nowcasts and very-short-range forecasts will become possible. However, this basic and extremely important concept hinges on learning how to integrate satellite data and other meteorological data into cohesive data sets based on the uniformly calibrated, single sensor advantage provided by the geostationary satellite.

3 Mesoscale applications using geostationary satellite imagery

3.1 A SEEMINGLY EASY NOWCAST AND VERY-SHORT-RANGE FORECAST SITUATION: FOG AND STRATUS

Let us examine what at first might appear to be a fairly simple problem: the evolution of early morning fog and stratus. Many questions can be asked concerning this situation and a related nowcast. For example: Why is it there? How extensive is it? Where are its boundaries? Is it moving, growing, or dissipating? How soon will it go away? Will it affect other weather development on local scales later in the day? What will be its effect on local temperatures throughout the day? The list of pertinent questions regarding the nowcast can grow almost indefinitely. It should come as no surprise that the questions concerning the extent of coverage and its boundaries can normally be answered only through use of satellite imagery. Perhaps more surprising, is that satellite is the best single source of data available to help answer most of the other questions listed above.

What are the effects of the fog and stratus? In the cloud-free areas, the energy of the sun may freely heat the ground and air, while the cloudy areas are kept several degrees cooler, due mainly to the higher albedo of the cloud. This cloud versus no cloud obviously affects surface temperature evolution; less obvious but no less real is its effect on afternoon cloudiness, and, under proper conditions, convective shower development as well as how rapidly the stable cloud will dissipate in various areas.

3.1.1 Dissipation

In two papers on radiationally induced fog and stratus dissipation, Gurka (1974, 1978a) showed dissipation to be a function of the time of year, reflected brightness of the fog and stratus, and the location of its boundaries. Dissipation was found to occur from the outside edges of the cloudy region inward, in part due to mixing as a result of differential heating along the cloud boundary. Additionally, brighter cloud areas were found to dissipate later than less bright regions. Using these facts, Gurka (1978b) developed a methodology for using information in visible satellite imagery to predict the time of fog dissipation at any point within the fog area. Operational utilization of methodology such as this awaits the technology of PROFS or other systems with similar very-short-range forecast goals.

3.1.2 Temperature

Many factors may influence surface air temperature. Among them are lapse rate, surface wind speed, surface characteristics and cloud cover. These factors have been discussed to some extent by Sutton (1953), Munn (1966), and Paltridge and Platt (1976).

Purdom (1973), Purdom and Gurka (1974), and Weiss and Purdom (1974) combined daytime satellite observations of cloud cover (pre-GOES) with surface temperature observations to aid in mesoscale surface temperature analyses. In those analyses they consistently found surface temperature to be several degrees cooler in the cloudy regions – this is as one might expect. Subsequent to that, using 1-km resolution infrared data from polar orbiting satellites, Carlson (1977) illustrated how that data could be used to study urban heat island patterns. Using computer-enhanced night-time 1-km resolution infrared data. Matson *et al.* (1978) and Matson and Legeckis (1980), have shown examples of temperature differences between rural and urban areas, as well as local differences within urban heat islands. Using poorer resolution GOES infrared data, Maddox and Reynolds (1980) have demonstrated a capability similar to those mentioned above in mapping extreme cold areas in Colorado. In addition, for the past few years, enhanced GOES infrared images have been used in the fruit forecast program in Florida (Miller, 1977).

These studies lay a foundation for nowcasting and very-short-range forecasting procedures. Within the fog and stratus region, the satellite-measured brightness should correlate inversely with the amount of sunlight reaching the ground, and thus indirectly with temperature. Furthermore, in the cloud-free areas the satellite infrared data can be used to infer surface temperature gradients. Proper combinations of these pieces of information (taking into account water vapour effects) with direct measurements of surface air temperature should allow for a more precise analysis of surface temperature. This type of initial analysis is a fundamental part of the nowcast. Temperature evolution over the next few hours is more complicated. However, knowledge of when fog and stratus will dissipate over a given area using the previously mentioned work of Gurka (1978b), as well as lapse rate, surface characteristics and expected cloud cover, all would certainly enter into the solution of the problem. For a striking example of the effect of transient cloud features on winds and temperatures see Maddox's (1977) study that combines GOES imagery and surface mesonet data.

3.1.3 Cloud and shower development

Given an area of early morning fog and stratus, where should one expect the first convective clouds and/or showers to form later in the day? It is well known that the static stability of the lower portion of the troposphere is in large part

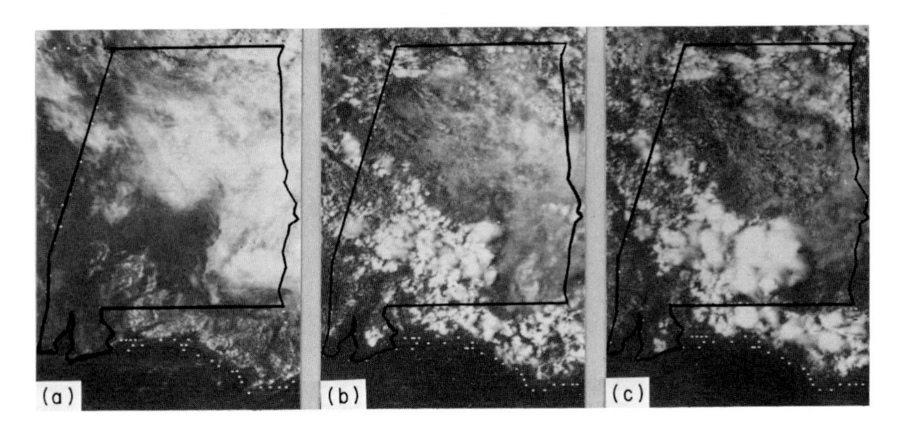

Fig. 1. GOES visible images for 27 May 1977 for (a) 1530, (b) 1830 and (c) 1930 GMT. The state shown is Alabama (southeast United States), which encompases an area of about 130 000 km^2. This series of images shows the dramatic effect early cloud cover can have on afternoon thunderstorm development. Note that the early clear region in southwest Alabama becomes filled with strong convection during the day, while the early cloudy region over the remainder of the state evolves into mostly clear skies. Also notice how the strongest activity later in the day develops in the "notch" of the clear region in south central Alabama, as one might expect from merging cloud breeze fronts.

controlled by differential heating (Dutton, 1976). This heating in turn affects the depth of the mixed layer, which partly determines whether or not convective clouds will form.

Purdom (1973), and Purdom and Gurka (1974) discussed the effects of early morning cloud cover on afternoon thunderstorm development in a synoptically weakly forced atmosphere. The situation was likened to that of the land-sea breeze (Haurwitz, 1947), with the first showers forming in the clear region near the boundary of the early morning cloud cover – a sort of cloud–breeze front. Additionally they found the slower heating rate in the early cloudy areas helped keep those regions free from convection for most of the day. Figure 1 is a good example of this phenomenon. This problem is more difficult to cope with than a land–sea breeze regime in that the early cloud boundary is continuously moving and dissipating, and thus the differential heating mechanism is continuously changing character. The effect of cloud boundaries on shower development is considered elsewhere in this volume by Carpenter (Chapter 4.2).

In a more stable atmosphere, similar results should be expected. In an example shown by Gustafson and Wasserman (1976) an early stratus region evolved into totally cloud-free skies later in the day, while, for one hundred or more kilometres on each side the sky was covered with cumulus. What makes this example particularly interesting is that the stratus area was moving during the day as it eroded, with the clear region that developed continuing to move in a similar fashion. This situation would have surely led to an incorrect very-short-range forecast for local cloud cover unless sequential satellite images had been available.

3.2 A MORE DIFFICULT SITUATION: CONVECTIVE STORMS

The convective storm is the cause of many weather-related events that dramatically affect life. Tornadoes, flash floods, downbursts and severe thunderstorms pose serious threats to life and property over many parts of the world. Although numerical weather prediction models employ various parameterization schemes to enter sub-grid scale feedbacks into the model, one of the major problems these schemes encounter is the handling of convection. This fact is certainly supported in recent studies of mesoscale convective complexes (Maddox, 1980) and of high plains severe thunderstorms (Doswell, 1980).

Basic to the very-short-range forecasting of the convective storm is understanding the mechanisms that lead to its development and evolution: this remains one of the most difficult problems in meteorology today. While it is generally agreed that the development of a convective storm or storm array depends on the interaction of meteorological fields ranging from the synoptic scale cyclone down to the cloud condensation nuclei (Lilly, 1975), little is known about why a particular storm forms and develops the way it does, or why a convective array behaves in a particular manner. Most of the available information about convective storm development and intensification is focused on either the large-scale conditions favourable for convective development (Miller, 1972), or on the individual convective storm (Browning, 1964; Fankhauser, 1971, 1976; Brandes, 1978). Indeed as Simpson, *et al.* (1974) point out, more is known about synoptic scale conditions favourable for convective development on the one hand and cloud microphysics on the other than is known about the mesoscale. The lack of information on the interaction between convective clouds and their mesoscale environment has been mainly due to a gap in meteorological observing capability in the days prior to the high resolution geostationary satellite.

On selected days during the past few convective seasons, the National Earth Satellite Service (NESS) has operated the GOES system in a special three-minute interval imaging mode. Radar PPI images over the convective areas were also collected on those days. This unique data set is allowing meteorologists for the first time to observe the development of deep convective storms with a spatial and temporal resolution compatible with the scale of the mechanisms responsible for their triggering. Films made from these data show that convective scale interaction is of primary importance in determining the evolution of deep convection. In fact, thunderstorm evolution that may appear random with radar is often observed as very well ordered when viewed with GOES imagery. This is because with satellite imagery we can observe all phases of deep cumulus developments, while with conventional radars we observe only the precipitation phase.

3.2.1 Mesoscale analysis: An initial step to understanding

An accurate and timely mesoscale surface analysis would be one of the most important tools available to the forecaster in detecting phenomena that will lead to convective development and intensification. Unfortunately, most detailed surface mesoscale analyses are done for post-storm research, require detailed observations from special mesoscale networks, and are too time consuming and involved for real-time forecasting. When applied in real-time situations they often fail because of the lack of a reporting station at the right place at the right time. Now, however, with GOES data, we have a "reporting station" every 1 km using the information in visible data, and every 8 km using infrared data.

Purdom (1976) pointed out that many of the mesoscale phenomena important in the initiation and maintenance of convection, such as the sea breeze (Pielke, 1974), dry lines (Rhea, 1966; Purdom, 1971), lake breeze (Lyons, 1966), areas of pre-squall line development (Miller, 1972), areas of convective cloud merger (Woodley and Sax, 1976), and mesoscale high-pressure systems (Fujita, 1963) which the forecaster previously tried to infer from macroscale patterns, are readily detectable in GOES imagery. Purdom (1973, 1974) showed how satellite data could be combined with conventional surface observations to construct an accurate and timely mesoscale surface analysis. Maddox (1977), in studying variability in surface mesonet data, demonstrated the important role high-resolution GOES imagery can play in the interpretation and analysis of meso-network data. An example of a mesoscale surface analysis and forecast using satellite data is shown in the example below.

3.2.2 The case of 26 May 1975

This case has previously been discussed by Purdom (1976, 1979). Using satellite imagery, Purdom (1973), showed the leading edge of a mesohigh appeared as an arc-shaped line of convective clouds moving outward from a thunderstorm area. Later, Purdom (1974) showed that the intersection of an "arc" cloud with another boundary marked a point with a high potential for severe weather. This correlated well with the work of Miller (1972), and later work of Maddox *et al.* (1980), on the importance of boundaries in severe thunderstorm and tornado development.

On 26 May 1975, a cold front trailed southwestward from northern Arkansas, becoming stationary across the Texas/Oklahoma border. At 2000 GMT (Fig. 2(a)) the cold front can be located by the convection along it in northern Arkansas. The stationary front may be located by combining the satellite

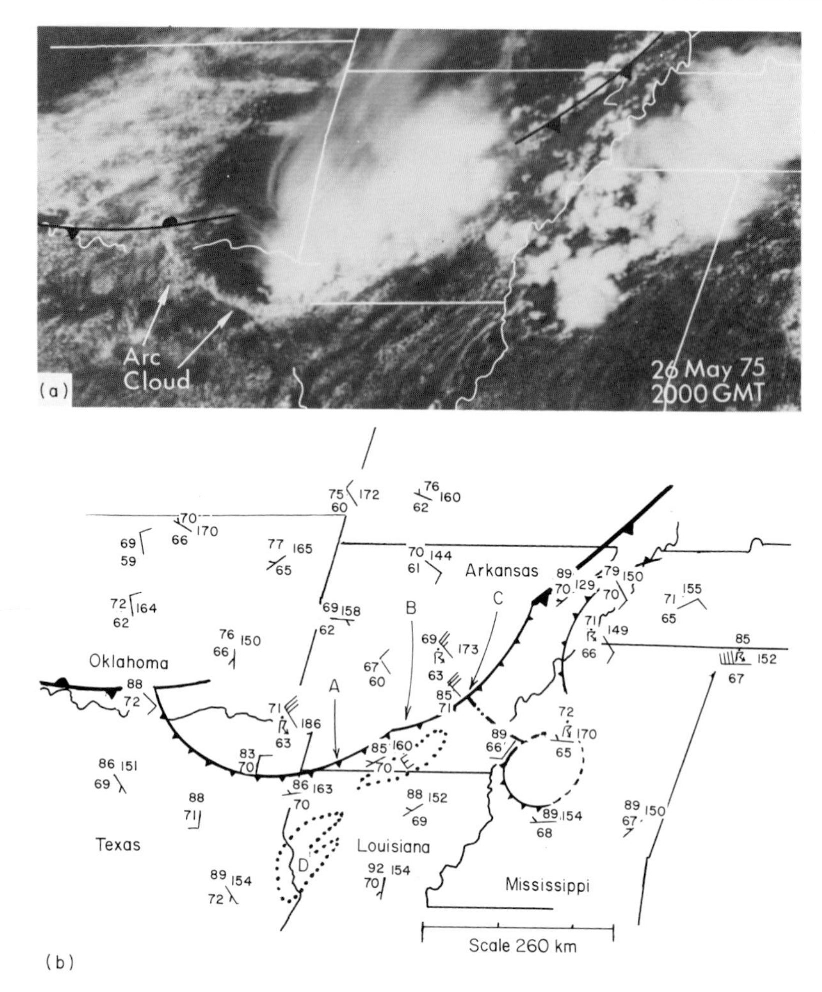
Arc
Cloud
(a)
26 May 75
2000 GMT
Oklahoma
Arkansas
Texas
Louisiana
Mississippi
A
B
C
D
E
Scale 260 km
(b)

Severe
Arc
(c)
26 May 75
2230 GMT

imagery (Fig. 2 (a)) with surface observations (Fig. 2 (b)) and noting the difference in cloud character to its north (stratiform) versus south (cumulus). An active mesoscale high-pressure system is moving through central to southwestern Arkansas, with its western boundary showing up as the arc cloud in northeast Texas. At this time no precipitation was detected by the Oklahoma City radar. Figure 2(c) shows the severe thunderstorm that developed where the arc cloud intersected the stationary front. It suggests that the precise point of thunderstorm formation could be identified earlier using satellite data than by using radar. In fact, the radar from Oklahoma City altogether failed to detect the arc cloud which served as the trigger mechanism.

Forecasting where the severe thunderstorm should develop on this day was actually among the easier of the convective forecast problems. Far more difficult to predict was the hour-by-hour behaviour of thunderstorm activity as the mesoscale high-pressure system pushed into northern Louisiana (Figs. 3a and 3b). This was discussed in Purdom (1979), and the presentation that follows is from that paper. It is a case of selective development. Inspection of Figs. 3a and 3b shows significant variations in thunderstorm activity as the arc moves through southern Arkansas into northern Louisiana and Mississippi. Why this variation? The clue to the answer lies in the variation in the cumulus field through which the arc moved. This variation in the cumulus field is a reflection of mesoscale variations in wet bulb potential temperature $(\theta)_w$ and vertical motion (w) ahead of the arc. θ_w and w should be larger in the cumulus areas than in the clear regions, making the cumulus-filled air more favourable for sustaining deep convection upon triggering (Purdom, 1979). This cumulus field variation is not detectable in conventional type mesoscale analyses; it is only obtained by careful analysis of satellite imagery.

In Fig. 2(a), there are thunderstorms all along the arc from A to C. A little over one and a half hours later, the situation was as shown in Fig. 3a namely: (1) thunderstorm activity continued along the western portion of the arc, at A, as it moved through what was previously a fairly homogenious cumulus field to

Fig. 2. (a) GOES visible image for 26 May 1975 at 2000 GMT with a resolution of 1 km. Note the arc cloud in northeast Texas and southern Arkansas. This arc played a significant role in the development of weather in the area later in the day.

(b) Surface data at 2000 GMT for 26 May 1975 with important satellite features from Fig. 2a. Note that the exact location of features such as fronts and mesohigh boundaries (arcs) may be readily extracted from the satellite imagery, both increasing the accuracy and reducing the time required for mesoscale analysis.

(c) GOES visible image for 26 May 1975 at 2230 GMT with a resolution of 1 km. Note the severe thunderstorm that has developed precisely at the intersection of the arc cloud and stationary front along the Oklahoma and Texas border. This location for preferred severe thunderstorm development was identifiable several hours prior to thunderstorm formation using GOES imagery. Other convection is discussed in connection with Figs 3a and 3b.

its south; (2) in the middle, at B, no thunderstorms formed as the arc moved into the clear area E; (3) strong thunderstorms developed at C as the arc intercepted developing convection to its east. About 2 hours later (Fig. 3b) intense convection was continuing at C; however, there had been a complete reversal in the type of convection at other places along the arc. Thunderstorm activity along segment A dissipated as the arc moved into the clear area at D, while along B to its east new thunderstorms formed as the arc moved into an area of preexisting cumulus cloudiness to the south of E.

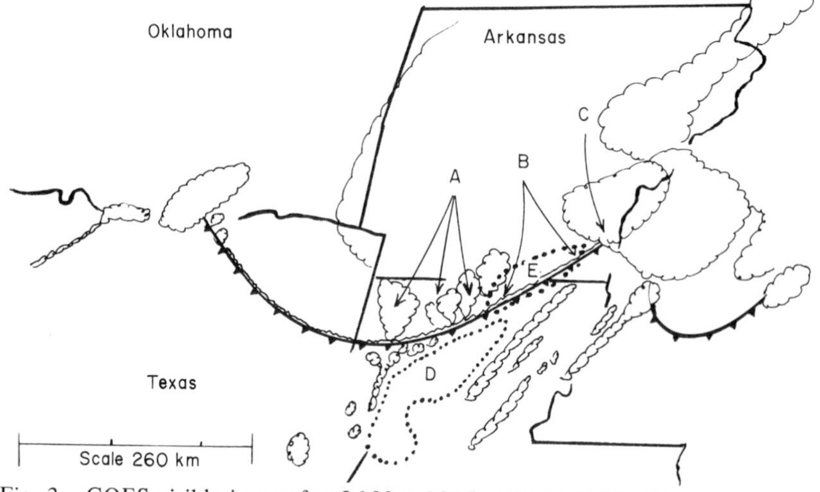

Fig. 3a. GOES visible image for 26 May 1975 at 2132 GMT with a resolution of 1 km, and hand drawing of important features in that image. Variations in convection along the arc are discussed in the body of the text.

3.2.3 Trigger mechanisms

As mentioned in section 3.2.1, many of the phenomena important in the initiation of convection are readily detectable in satellite imagery. Generally, organized convergence lines that might trigger strong convection, such as some fronts, pre-frontal squall lines or dry lines, can be detected in satellite imagery prior to deep convection developing on them and being detected by radar. An example of dry line development is shown in Figs. 4a and 4b. In this case a dry

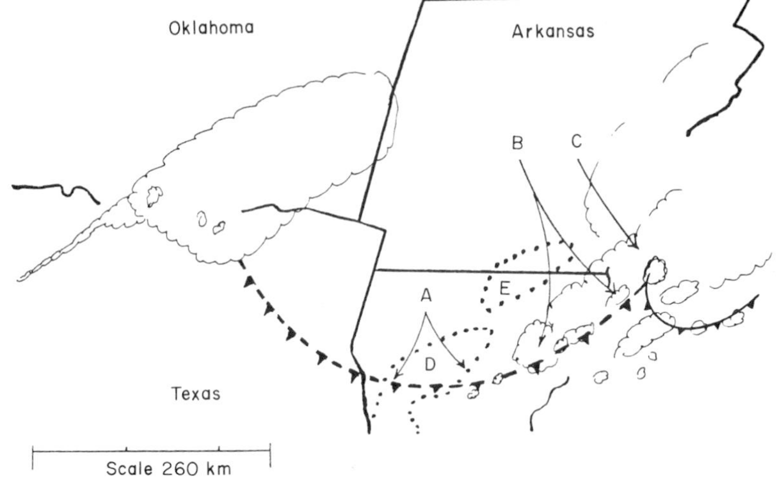

Fig. 3b. GOES visible image for 26 May 1975 at 2347 GMT with a resolution of 1 km, and hand drawing of important features in that image. For discussion see text.

Fig. 4a. GOES visible image for 14 June, 1976 at 2000 GMT with a resolution of 1 km. This image shows a narrow dry line, extending from central Kansas into eastern Nebraska and South Dakota, prior to the development of severe convection along it.

line which extended from eastern South Dakota and Nebraska into central Kansas at 2000 GMT (Fig. 4a) developed into a line of severe thunderstorms as it moved into Minnesota and Iowa by 2300 GMT (Fig. 4b).

Near the time of the image shown in Fig. 4b, large hail and funnel clouds were reported in eastern South Dakota, a tornado injured six people in Minnesota, and a weak tornado was reported in Iowa. Later in the evening, six people were injured and one was killed by tornado activity in Minnesota, severe

Fig. 4b. GOES visible image for 14 June, 1976 at 2300 GMT with a resolution of 1 km. This image shows the severe convection that has developed along the dry line as it pushed into Minnesota and Iowa. For information on the character of the severe weather see the text.

winds were reported in Iowa, and the thunderstorms in Central Kansas (just beginning to develop in Fig. 4b) had numerous funnel reports, one confirmed tornado and several reports of hail.

Not all convective triggering mechanisms have received as much attention as those which cause severe thunderstorms. This is especially true of thunderstorm development in a synoptically weakly forced atmosphere. In a recently completed study using GOES satellite data, Purdom and Marcus (1982) classified convective

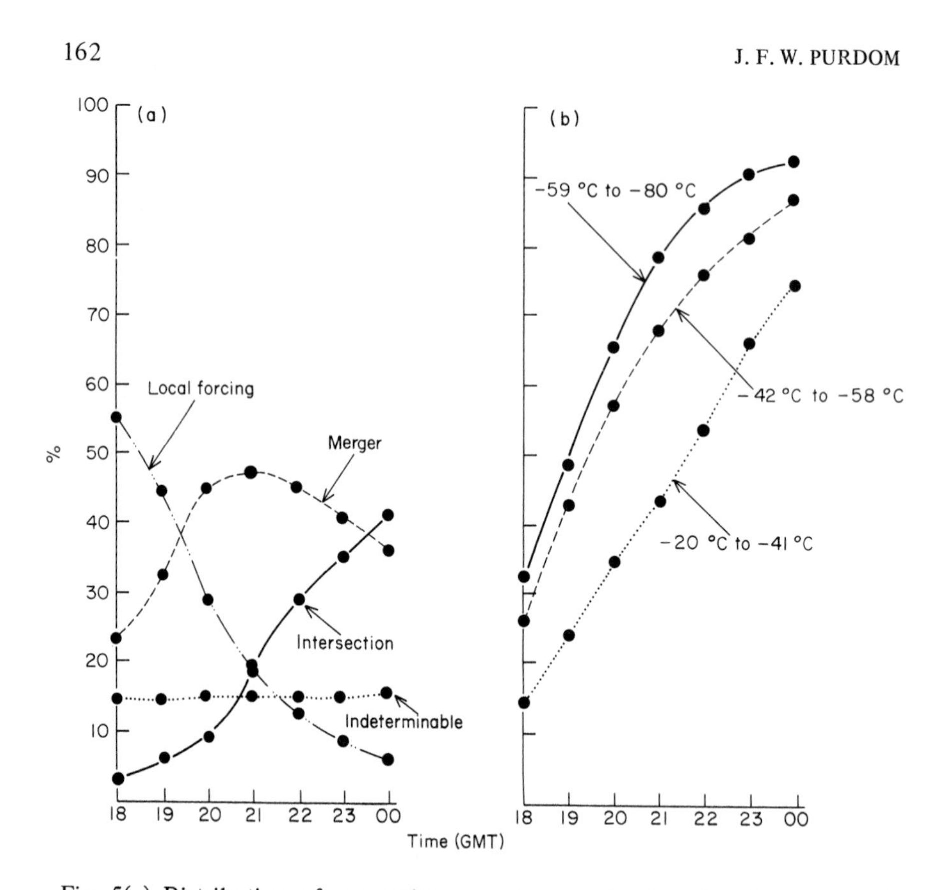

Fig. 5(a) Distribution of convective generation mechanisms for all storms with tops colder than $-20°C$ versus time for the summer of 1979 over the South-east United States. The sample contained over 9850 storms.

(b) Percentage of storms in various temperature ranges due to arc cloud intersections and mergers for a given hour (assuming indeterminable storms in Fig. 4a are equally distributed). For example: of those storms with tops between $-59°C$ and $-80°C$ at 0000 GMT, 93% are due to merger and intersection of arc clouds.

development over the southeast United States (an area of approximately 7.8×10^5 km^2) for the summer of 1979. Over 9850 convective storms were classified with respect to storm generation mechanism and intensity from 1800 to 0000 GMT (noon to 6 pm local). Intensity was classified using GOES infrared temperature, with all storm tops colder than $-20°C$ being classified. Generation mechanisms were classified into four categories: (1) merger, i.e. development on an arc cloud as it moved into a non-precipitating cumulus region; (2) intersection, i.e. development where two arc clouds came into contact; (3) local forcing, i.e. development due to some local mechanism not involving arc clouds;

(4) indeterminable i.e. development whose generation mechanism could not clearly be determined, such as new storms from beneath a cirrus deck. Figure 5 summarizes the results for the entire study period. Those results show quantitatively that, while early in the day local forcing dominates as the convective generation mechanism, later in the day when the most intense convection has developed, the dominant generation mechanisms are mergers and intersections of arc clouds. Intersections and mergers were found to occur between 25 and 300 km from the storm that produced the arc, with the majority occurring at distances between 50 and 150 km on a time scale of one to three hours. The motion of an arc may be computed using image pairs, while the evolution of the cumulus field in advance of the arc may be monitored as in Figs 2(a) and 3b. Information such as this, when presented rapidly and in the proper format to a forecaster, is very important in the very-short-range forecasting of thunderstorm development and evolution.

4 Combining satellite and radar data

The question of how best to use satellite and radar information together for nowcast purposes is still far from resolved. Most commonly, these two data sets have been analysed separately, then mentally merged by the meteorologists. Recently, Reynolds and Smith (1979) have used the Colorado State University All Digital Video Imaging System for Atmospheric Research, ADVISAR (Smith *et al.*, 1978), to combine digital satellite and radar data into composite displays. They pointed out the potential value of such composites as aids to both the researcher in case study analysis and the forecaster in real-time decision making. The work of Reynolds and Smith is encouraging, with the potential for new developments in this type of visual presentation being enormous.

Little work has yet been done in extracting quantitative information about mesoscale convective features and individual thunderstorms by combining digital information from satellite and radar. Papers by Adler and Fenn (1979) and Negri and Adler (1981), in which thunderstorm intensity determined from 5-minute interval GOES data are examined in relation to simultaneous radar measurements, are certainly initial steps in this direction. However, these efforts are more of a comparison of the two different data types (certainly necessary) rather than a combination. This area has significant potential for further development. The direction this development will take is an open question.

Both satellite and radar provide independent measures of thunderstorm intensity. Satellite data provides such information as cloud top mean vertical growth and anvil expansion rates, while radar data provides information about reflectivity, volumetric echo properties and their changes in time. Perhaps combining these pieces of information in a simple diagnostic cloud model will lead to an intensity measure for thunderstorms. This information could then be added to the satellite image in a manner similar to that used by Reynolds and Smith.

Before the convective storms have developed, radar is less useful and satellite imagery and surface observations should be combined instead. This would allow the detection of trigger mechanisms and the delineation of areas with deep convective potential from those without it. (Radar has shown some promise in this area, and one should be aware of the works of Harrold and Browning, 1971, and Doviak, 1981). Next, as suggested in the above paragraph, precipitation areas and their intensities should be determined using satellite and radar data. As a final product, the information on intensity and area could be merged with the satellite image and surface observations. Using a system such as ADVISAR, different colours could be used in the convective areas to indicate storm intensity and severity.

5 Conclusion

Geostationary satellite data is likely to become one of the major footings upon which mesoscale forecasting programs of the future are based. By combining satellite data with more conventional data, such as radar and surface observations, many of the features important in mesoscale weather development and evolution may be better analysed and understood. This better analysis and understanding of mesoscale processes is necessary if very-short-range forcasting is to be successful.

References

Adler, R. F. and Fenn, D. D., (1979) Thunderstorm intensity as determined from satellite data. *J. appl. Met.* **18**, 502–517.

Beran, D. W. (ed.) (1980). PROFS proposed policy, mission, goals, and organizational structure. NOAA Technical Memorandum ERL PROFS-2, U.S. Department of Commerce, Boulder, Colorado, 51pp.

Brandes, E. A. (1978). Mesocyclone evolution and tornadogenesis: some observations. *Mon. Weath. Rev.* **106**, 995–1101.

Browning, K. A. (1964). Airflow and precipitation trajectories within severe local storms which travel to the right of the winds. *J. atmos. Sci.* **21**, 634–639.

Carlson, T. N., Augustine, J. N. and Boland, F.E., (1977). Potential application of satellite temperature measurements in the analysis of land use over urban areas. *Bull. Am. met. Soc.* **58**, 1301–1303.

Doviak, F. J. (1981). Doppler weather radar for forecast and warnings. Introductory Geoscience and Remote Sensing Symposium (IGARSS-81), pp. 152–157. IEEE, Washington, D.C.

Dutton, J. A. (1976) "The Ceaseless Wind", 579 pp. McGraw-Hill, New York.

Fankhauser, J. (1971). Thunderstorm-environment determined from aircraft and radar observations. *Mon. Weath. Rev.* **99**, 171–192.

Fankhauser, J., (1976). Structure of an evolving hail storm, part II: Thermodynamic structure and airflow in the near environment. *Mon. Weath. Rev.* **104**, 576–587.

Fujita, T. T. (1963). Analytical mesometeorology: A review. Met. Monogr. No. 27, 77–128.
Gurka, J. J. (1974). Using satellite data for forecasting fog and stratus dissipation. Preprints, 5th Conference on Weather Forecasting and Analysis (St Louis, Missouri), pp. 54–57. American Meteorological Society, Boston, Massachusetts.
Gurka, J. J. (1978a). The role of inward mixing in the dissipation of fog and stratus. *Mon. Weath. Rev.* **106**, 1633–1635.
Gurka, J. J. (1978b). The use of enhanced visible imagery for predicting the time of fog dissipation. Preprints Conference on Weather Forecasting and Analysis and Aviation Meteorology (Silver Spring, Maryland), American Meteorological Society, Boston, Massachusetts.
Gustafson, A. V. and Wasserman, S. E. (1976). Use of satellite information in observing and forecasting for dissipation and cloud formation. *Mon. Weath. Rev.* **104**, 323–324.
Harrold, T. W. and Browning, K. A. (1971). Identification of preferred areas of shower development by means of high power radar. *Q. J. R. met. Soc.* **97**, 330–339.
Haurwitz, B. (1947). Comments on the sea breeze circulation. *J. Met.* **4**, 1–8.
Lilly, K. K. (ed.) (1977). Project SESAME: Planning documentation volume. NOAA/ERL, U.S. Department of Commerce, Boulder, Colorado, 308pp.
Ludwig, G. H. (1974). The NOAA operational environmental satellite system; status and plans. Preprints, 6th Conference on Aerospace and Aeronautical Meteorology (El Paso, Texas) pp. 137–145. American Meteorological Society, Boston, Massachusetts.
Lyons, W. A. (1966). Some effects of Lake Michigan upon squall lines and summertime convection. SMRP Research Paper 57, Department of Geophysical Science, The University of Chicago, 22 pp.
Maddox, R. A. (1977). Meso-B scale features observed in surface network and satellite data. *Mon. Weath. Rev.* **105**, 1056–1059.
Maddox, R. A. (1980). Mesoscale convective complexes. *Bull. Am. met. Soc.* **61**, 1374–1387.
Maddox, R. A. and Reynolds, D. W. (1980). GOES satellite data maps areas of extreme cold in Colorado. *Mon. Weath. Rev.* **108**, 116–118.
Maddox, R. A., Hoxit, L. R. and Chappell, C. F., (1980). A study of tornadic thunderstorm interactions with thermal boundaries. *Mon. Weath. Rev.* **108**, 322–336.
Matson, M. and Legeckis, R. V. (1980). Urban heat islands detected by satellite. *Bull. Am. met. Soc.* **61**, 212.
Matson, M., McClain, E. P., McGinnis, D. F. Jr. and Pritchard, J. A., (1978). Satellite detection of urban heat islands. *Mon. Weath. Rev.* **106**, 1725–1734.
Miller, R. C. (1972). Notes on analysis and severe-storm forecasting procedures of the Air Force Global Weather Central. AWS Technical Report 200 (rev), Air Weather Service (MAC), U.S. Air Force, 190 pp.
Miller, D., Waters III, M. P., Tarpley, J. D., Green, R. N., and Dismachek, D. C., (1977). Potential applications of digital, visible, and infrared data from geostationary environmental satellites. Proceedings 11th International Symposium on Remote Sensing of Environment, Environmental Research Institute of Michigan, (Ann Arbor, Michigan).
Mosher, F. R. (1980). Compatibility of Cloud tracked winds from the United States, European and Japanese Geostationary Satellites. *Advances in Space Research,* **I**, 139–146.

Munn, R. E. (1966) "Descriptive Micrometeorology", 245 pp. Academic Press, New York and London.

Negri, A. J. and Adler, R. F., (1981). Relation of satellite-based thunderstorm intensity to radar-estimated rainfall. *J. appl. Met.* **20**, 66–78.

Paltridge, G. W. and Platt, C. M. R., (1976). "Radiative Processes in Meteorology and Climatology", 318 pp. Elsevier Scientific Publishing Co., New York.

Pielke, R. (1974). A three-dimensional numerical model of the sea breezes over south Florida. *Mon. Weath. Rev.* **102**, 115–139.

Purdom, J. F. W. (1971). Satellite imagery and severe weather warnings. Preprints 7th Conference on Severe Local Storms, (Kansas City, Missouri) pp. 120–127. American Meteorological Society, Boston, Massachusetts.

Purdom, J. F. W. (1973). Meso-highs and satellite imagery. *Mon. Weath. Rev.* **101**, 180–181.

Purdom, J. F. W. (1974). Satellite imagery applied to the mesoscale surface analysis and forecast. Preprints, 5th Conference on Weather Forecasting and Analysis (St. Louis, Missouri), pp. 63–68. American Meteorological Society, Boston, Massachusetts.

Purdom, J. F. W. (1976). Some uses of high-resolution GOES imagery in the mesoscale forecasting of convection and its behavior. *Mon. Weath. Rev.* **104**, 1474–1483.

Purdom, J. F. W. (1979). The development and evolution of deep convection. Preprints 11th Conference on Severe Local Storms (Kansas City, Missouri), pp. 143–150. American Meteorological Society, Boston, Massachusetts.

Purdom, J. F. W. and Gurka, J. G. (1974). The Effect of early morning cloud cover on afternoon thunderstorm development. Preprints, 5th Conference on Weather Forecasting and Analysis (St. Louis, Missouri), pp. 58–60. American Meteorological Society, Boston, Massachusetts.

Purdom, J. F. W. and Marcus, K. (1982). Thunderstorm trigger mechanisms over the southeast United States. Preprints 12th Conference on Severe Local Storms (San Antonio, Texas), pp. 487–488. American Meteorological Society, Boston, Massachusetts.

Reynolds, D. W. and Smith, E. A. (1979). Detailed analysis of composited digital radar and satellite data. *Bull. Am. met. Soc.* **60**, 1024–1037.

Rhea, J. O. (1966). A study of thunderstorm formation along dry lines. *J. appl. Met.* **5**, 58–63.

Simpson, J. and Dennis, A. S. (1974). Cumulus clouds and their modification. *In* "Weather Modification" (W. N. Hess, Ed.), pp. 229–280. Wiley, New York.

Smith, E. A., Brubaker, T. A. and Vonder Haar, T. H. (1978). All digital video imaging system for atmospheric research (ADVISAR). Technical Report, Department of Atmospheric Science, Colorado State University, Fort Collins, Colorado, 8 pp.

Smith, W. L., Soumi, V. E., Menzel, W. P., Woolf, H. M., Sromousky, L. A., Revercomb, H. E., Hayden, C. M., Erickson, D. N. and Mosher, F. F. (1981). First sounding results from VAS-D. *Bull. Am. met. Soc.* **62**, 232–236.

Sutton, O. G. (1953). "Micrometeorology", 333 pp. McGraw-Hill, New York.

Weiss, C. E. and Purdom, J. F. W. (1974). The effect of early morning cloud cover on squall line activity. *Mon. Weath. Rev.* **102**, 400–401.

Woodley, W. L. and Sax, R. I. (1976). The Florida area cumulus experiment: rationale, design, procedures, results, and future course. NOAA Technical Report ERL 354–WMPO 6, 204 pp.

3.2

Automated Techniques for the Analysis of Satellite Cloud Imagery

ERIK LILJAS

1 Introduction

Remote sensors are capable of providing many of the mesoscale observations required to identify and track mesoscale weather systems. These observations can be put to good use independently of a numerical model. Given frequently updated descriptions of the current weather the procedure is to extrapolate the observed motion and changes in intensity. Extrapolation of weather systems can be done either with manual or automatic methods (Austin and Bellon, 1982). In their simplest form these methods do not take into account future growth or decay and are not suited to forecasting more than 3–6 hours ahead on the mesoscale. Three main factors influence the growth and decay of mesoscale systems:

i. Stability of the atmosphere.
ii. Stage of development of the weather system or phenomenon.
iii. Local effects due to mountains and land-water boundaries, cooling or warming from the underlying surface etc. Small weather systems are highly dependent on local effects. These effects are therefore very important in forecasting mesoscale systems.

The forecaster's role in taking these factors into account can be of great importance in forecasts 0–12 hours ahead.

As discussed by Purdom (1982), cloud types, cloud patterns and their evolution are some of the best indicators of atmospheric processes, ranging from synoptic-scale disturbances to small-scale convection. One of the principal missions of weather satellites is to provide this information in picture form. The latest generation of weather satellites gives data with high spatial and cloud top temperature resolution. Unfortunately a great deal of essential information is lost in the widely used quick-look visible and infrared pictures covering large areas and produced in black and white. The lack of adequate automated procedures for analysing and displaying satellite imagery is one major obstacle preventing wider application of satellite imagery.

In this chapter the results of a recent Swedish project (Liljas, 1981) to improve the utility of satellite imagery are reported. The objectives of the project were:

1. To classify different cloud types in terms of the satellite imagery and to present the analysed cloud patterns on a colour video display.
2. To derive a qualitative measure of precipitation intensity from the cloud imagery.
3. To reduce the very great volume of data in order to enable quick transmission to remotely located weather services.

2 An automated multispectral cloud and precipitation classification technique for the analysis of satellite imagery

This study was carried out for TIROS-N type polar orbiting satellites. These satellites provide digital data from the Advanced Very High Resolution Radiometer, AVHRR. Data were used from three spectral channels as follows:

VIS (visible)	ch. 1	0.55 – 0.9 μm 0.58 – 0.68 μm	(TIROS–N) (NOAA–6, –7)
NIR (near infrared)	ch. 2	0.725 – 1.1 μm	
IR (infrared)	ch. 4	10.5 – 11.5 μm	

AVHRR has a spatial resolution of 1.1 km at the subsatellite point and the radiational intensity is given in 2^{10} digital counts for each channel. The technique used is generally similar to that previously used to derive certain parameters from LANDSAT imagery. The method, involving a cloud brightness and texture discriminant analysis technique, resembles that proposed by Barrett and Grant (1978). The basis of the technique is that different cloud types, land and water surfaces have different radiational properties in different parts of the electromagnetic spectrum. Accordingly the satellite images in the three channels have been compared with ground-based observations of clouds using an interactive image analysis system. In this way signatures were derived of different cloud types within boxes in three-dimensional intensity space. The results of the analysis are presented in Figs 1, 2 and 3.

The boxes in the figures separate six important cloud types: cumulonimbus, nimbostratus, cirrostratus, cumulus congestus, stratocumulus and stratus or fog. Cumulus humilis has the same intensity classification as fog or stratocumulus, and thick fog as stratocumulus. Separation of fog, stratocumulus and small cumulus

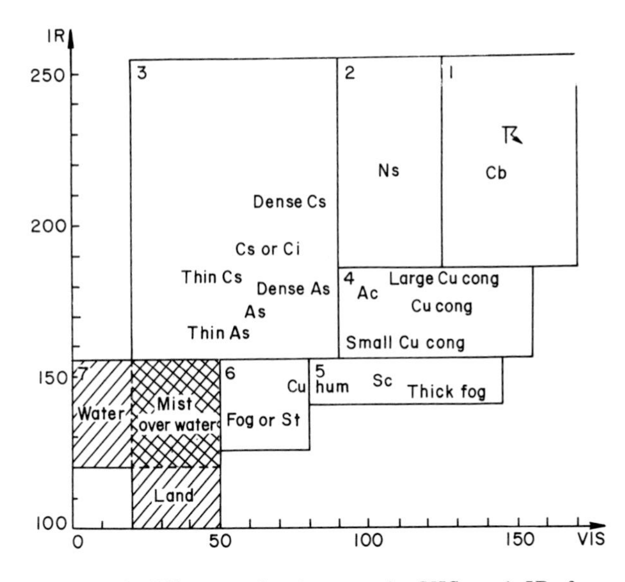

Fig. 1. Separation of different cloud types in VIS and IR from TIROS–N AVHRR-data. Vertical axis is IR temperature in digital counts and horizontal axis is cloud reflectivity. Six boxes separate the main cloud categories. Land and water overlap. Sun elevation 45°.

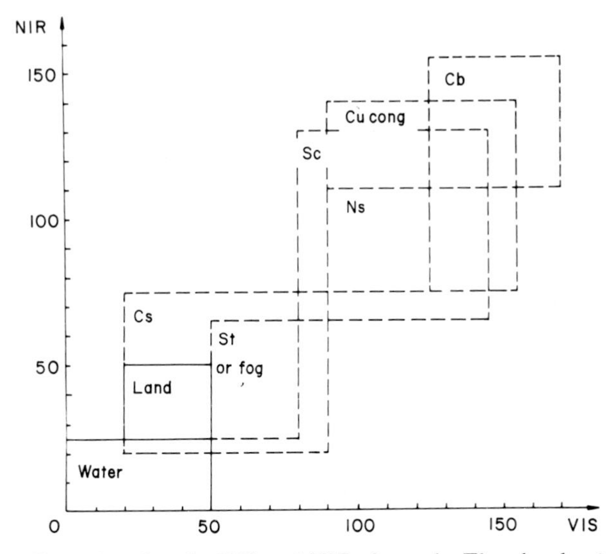

Fig. 2. As for Fig. 1 but for the VIS and NIR channels. The cloud categories overlap and the NIR channel does not improve the cloud separation compared with that achievable with VIS and IR alone, but water and land surfaces are separated.

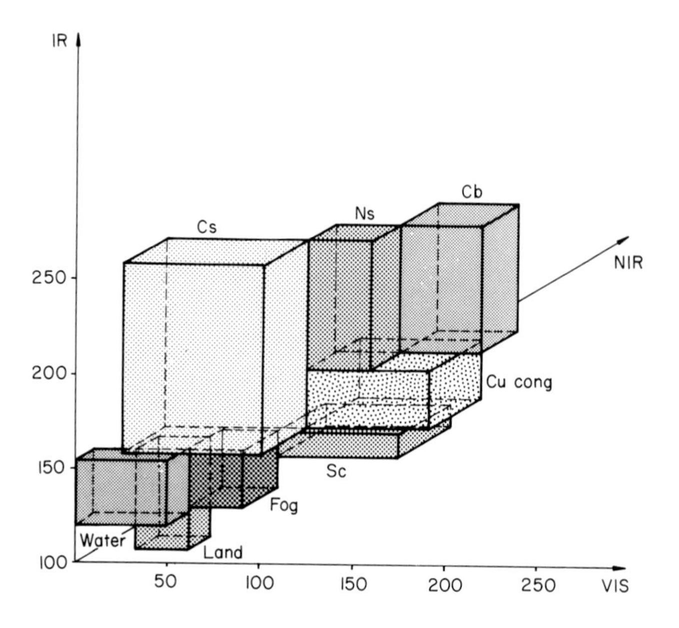

Fig. 3. The classification in a three-dimensional intensity space. Slightly different classifications can be built up suited to various weather types.

can, however, be achieved manually from the texture of high-resolution images. Cumulus humilis clouds are seen as small cloud elements often in lines or with cellular patterns. Stratocumulus has a weak cell structure. Fog or stratus has mostly a smooth texture with sharp boundaries. Except for stratus, fog and stratocumulus, the intensity signatures from well-defined cloud types do not overlap. In practice, however, there exist hybrids between different cloud types. For example, when there is a layer of cirrostratus or cirrus over low level clouds, the classification can be misleading.

The classification illustrated in Figs. 1, 2 and 3 represents TIROS-N AVHRR-data with a sun elevation angle of 45°. The reflected radiance in visible light and the near infrared will vary as the illumination changes with changing solar elevation. We thus require a more conservative property to associate with clouds and terrestrial background. One way would be to normalize the reflected data to zenith angle by dividing by the sine of the solar elevation angle. However, for a given cloud the reflectivity itself is also a function of the elevation angle of the sun (Muench and Keegan, 1979) and so instead of normalizing the imagery we have developed a set of classifications for different sun elevations.

In the study by Liljas (1981), several weather situations from a summer period were tested with the same threshold values in the classification. The images were presented on a colour TV screen and compared with ground

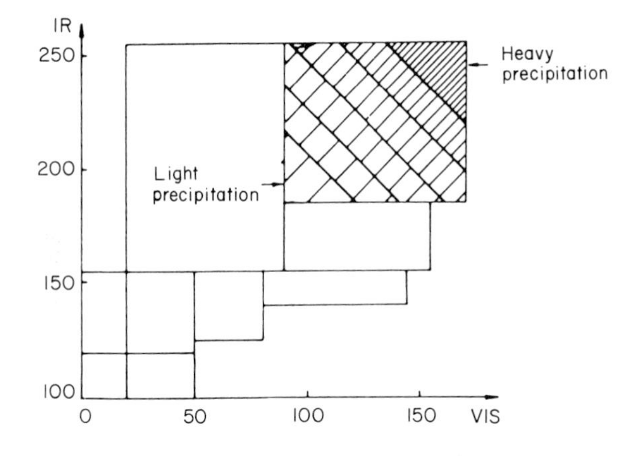

Fig. 4. Qualitative indication of precipitation intensities derived from AVHRR data. Vertical axis is IR temperature in digital counts and horizontal axis is cloud reflectivity. Very bright and cold tops give the highest rate.

observations. The method was found to give a correct indication of cloud types and, compared with synoptic observations, much better information concerning cloud distribution. In an operational system the classification scheme can be built up with "boxes" suited to different weather types. For instance a threshold can divide cumulonimbus into two parts, one with high probability of thunder, one with low probability. In a weather situation with only convective clouds, these can form the different boxes. Separation of different VIS-brightness values can also be provided in order to help forecast the dissipation of morning stratus or fog (Gurka, 1978).

Precipitation is often the most important parameter to identify and forecast. Interesting work within the area of rainfall estimation using satellite imagery has been described by Lovejoy and Austin (1979) and others. Their technique combines geostationary satellite visible and infrared data, using radar data to calibrate it. Since the method is indirect it provides probability of rain contours. The objective of the Swedish study was to apply a similar approach using real time TIROS-N type data, but without the radar to calibrate it. It was developed for both frontal and convective precipitation. The method starts with the classification of cloud types into cumulonimbus and nimbostratus as seen in Fig. 1. This region of VIS-IR space is sliced as shown in Fig. 4. As found by Lovejoy and Austin, clouds with very cold tops give the heaviest precipitation. A requirement for precipitation is that the cloud top temperature should be colder than $-12°C$.

The indicated precipitation has been compared with the available synoptic observations, airfield observations and four recording rain gauges. The method has been found to give good information on the distribution of precipitation. It gives an indication of the relative intensity of precipitation both in fronts and in showers. It has also been possible to detect intense precipitation in mesoscale cloud clusters between the synoptic ground stations.

3 The application of classified AVHRR data from TIROS-N

3.1 GENERAL RESULTS

The cloud classified images provide an opportunity for identifying different classes of mesoscale cloud systems so that, if the forecaster is familiar with the topography and its effects, he can determine if and how the cloud systems are constrained by surface effects. The colour images also give a good indication of the cloud structure in various systems, their stage of development and also the stability of the atmosphere. Stratocumulus, small cumulus elements and also daytime fog or stratus indicate stable conditions in the lower troposphere. Groups of large cumulus elements in the morning indicate humid and unstable conditions, and are often the early stage of a mesoscale system leading to cumulonimbus and thunder. Cumulus growing into cumulonimbus shows where a mesoscale convective systems is under development; widespread cirrostratus and cirrus is symptomatic of decay. A cloud system consisting of mainly cirrostratus or altocumulus generated from cumulonimbus is mostly inactive or decaying. In the case of baroclinic instability the early stages of frontal development can often be recognized in the patterns of deep stratiform clouds. Frontal bands consisting of broken altocumulus indicate a late stage of development. Intense weather systems are recognized as consisting of thick clouds with high tops. In the images they may be depicted by a red colour as a warning sign. The best evidence of growth or decay is obtained from successive images with the same classification scheme.

3.2 A CASE STUDY

Figure 5(a) shows a classified AVHRR colour image reproduced in black and white. Because the different cloud types are difficult to distinguish in black and white, the boundaries have been sketched in Fig. 5(b).

A frontal zone is oriented from A to B in Fig. 5(a). The left (western) part of the area is covered by cold air and the right (eastern) part by a warm moist southerly airstream. The ground observations in Fig. 6 confirm most of the cloud classification in Fig. 5(a) and (b).

The Roman numerals I–X in Fig. 5(a) represent different cloud systems identified in the satellite image. They may be summarized as follows:

I. Mainly nimbostratus with embedded convection associated with an active cold front.
II. Stratocumulus indicative of subsidence and stable conditions. No convection or development of rain-producing clouds is expected in this area.
III. An area of small and closely spaced cumulus cells associated with almost the same stability conditions as cloud system II.
IV, V and VI. Mesoscale thunderstorm systems developing in the stationary frontal zone. The thunderstorms, but not the mesoscale systems, are revealed by the synoptic ground stations. As the systems move northwards they redevelop at their southern edges where cumulus are seen growing into cumulonimbus. The systems decaying on their northern sides are characterized by high-level anvil outflows. The derived precipitation image indicated very heavy precipitation at the southern edges of systems V and VI, in association with the newly developed bands of convective cells.
VII. Part of a very deep stratiform cloud system, analysed as cumulonimbus because of its very dense nature and cold cloud top. The inferred precipitation intensity is high. The system is associated with a developing low on a frontal zone.
VIII, IX and X. Groups of cumulonimbus cells developing on the coast at end of cumulus lines where sea-breezes are forcing deep convection. These cells are terrain-induced but they sometimes cross the Baltic in the late evening when the sea-breeze fades out. The other systems in the image are more or less synoptically forced mesoscale systems and are advected with the midtropospheric winds.

4 Concluding remarks

Classified satellite images are not only valuable diagnostic tools for the forecaster, but they also imply a very great reduction in the volume of satellite data. The original image with three channels, 2^{10} intensity levels and covering an area of more than 2000 × 2000 pixels, will be reduced to 2^3 or 2^4 intensity levels (cloud types or precipitation intensities) and about 500 × 500 pixels. This new image can be transmitted via a telephone line within minutes. In the system discussed by Bodin (1982) the latest images will be stored so that movement and development of the weather systems can be easily determined by replaying successive images. The video system of the weather service ought to have capabilities for overlaying radar data and conventional fields of geopotential, air temperature, moisture and wind analysis to aid in the subjective interpretation of cloud patterns especially to determine whether they are likely to develop or decay.

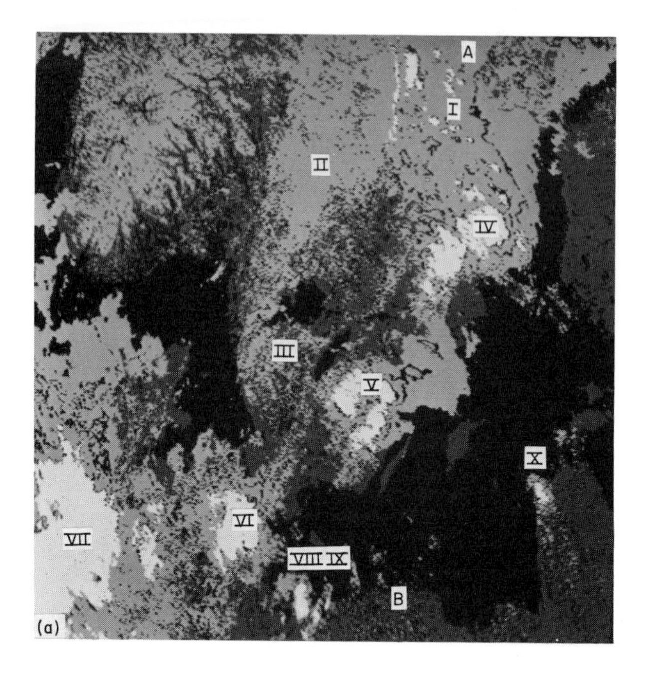
A
I
II
IV
III
V
X
VI
VII
VIII
IX
B
(a)

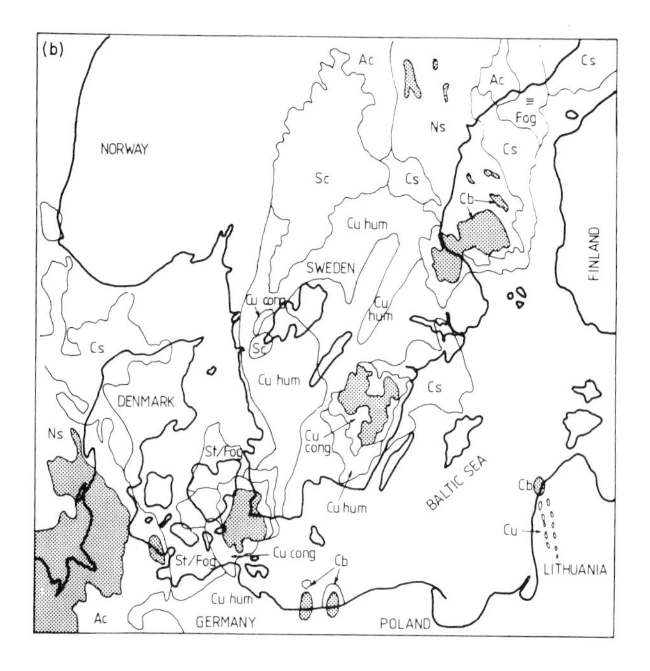
(b)
Ac
Cs
Ac
Fog
Ns
NORWAY
Cs
Sc
Cs
Cb
Cu hum
FINLAND
SWEDEN
Cu cong
Cu hum
Sc
Cs
Cu hum
Cs
DENMARK
Cu cong
Ns
St/Fog
BALTIC SEA
Cb
Cu hum
Cu
St/Fog
Cu cong
Cb
LITHUANIA
Cu hum
Ac
GERMANY
POLAND

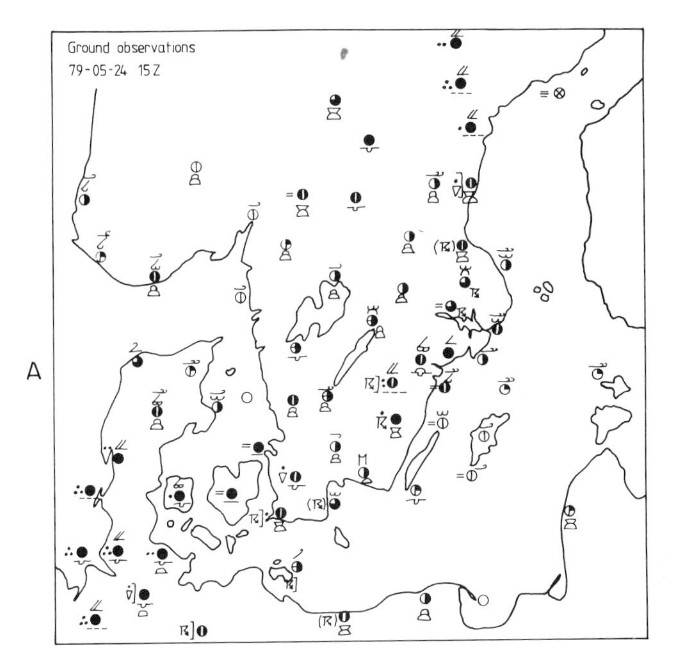

Fig. 6. Cloud and weather observed from the ground, at 1500 GMT. From the synoptic observations about $1\frac{1}{2}$ hours after the satellite passage.

References

Austin, G. L. and Bellon, A. (1982). This volume, pp. 177–190.

Barrett, E. C. and Grant, C. K. (1978). An appraisal of Landsat 2 cloud imagery and its implications for the design of future meteorological observing system. *J. Br. interplanet. Soc.* **31**, 3–10.

Bodin, S. (1982). This volume, pp. 25–36.

Gurka, J. J. (1978). The use of enhanced visible imagery for predicting the time of fog dissipation. Preprints, Conference on Weather Forecasting and Analysis and Aviation Meteorology (Silver Spring, Maryland), pp. 343–346. American Meteorological Society, Boston, Massachusetts.

Fig. 5 (*opposite*). (a) Classified AVHRR imagery from TIROS-N, 1310 GMT, 24 May 1979. The image covers an area of about 1200 × 1200 km over southern Scandinavia and the Baltic Sea. The original image was a colour video display, part of which is reproduced in the Frontispiece. The grey tones in (a) do not do justice to the easily identified cloud types and land-sea boundaries in the original image.

(b) Some of the different cloud types that can be identified in the original colour picture corresponding to (a). The shaded areas are classified as cumulonimbus.

Liljas, E. (1981). Analysis of clouds and precipitation through automated classification of AVHRR data. SMHI Reports RMK, **32**, 1–25.

Lovejoy, S. and Austin, G. L. (1979). The delineation of rain areas from visible and IR satellite data for GATE and mid-latitudes. *Atmosphere-Ocean* **17**, 77–92.

Muench, H. S. and Keegan, T. J. (1979). Development of techniques to specify cloudiness and rainfall rate using GOES imagery data. Environmental Research papers, No. 681, pp. 7–45. AFGL-TR-79-0255 Air Force Geophysical Laboratory, Hanscom Air Force Base, Massachusetts.

Purdom, J. F. W. (1982). This volume, pp. 149–166.

3.3

Very-short-range Forecasting of Precipitation by the Objective Extrapolation of Radar and Satellite Data

G. L. AUSTIN and A. BELLON

1 Introduction

The idea of using the extrapolation of radar maps of precipitation for preparing very-short-range weather forecasts is not new. Prior to the advent of digital radar data, simple extrapolation techniques which did not require computer hardware were devised by Ligda (1953), Russo and Bowne (1961), Boucher (1961) and Kessler (1961). Boucher (1963) classified precipitation echoes into three distinct categories for forecast purposes: squall lines, precipitation edges and amorphous fields, with the last exhibiting the highest predictability. Noel and Fleisher (1960) and Hilst and Russo (1960) pioneered the use of the maximum cross-correlation coefficient as the objective extrapolation predictor in forecasting precipitation. They also compared the forecast precipitation distribution with the actual precipitation pattern by means of an additional correlation technique.

Kessler and Russo (1963) and Kessler (1966) returned to a mathematical treatment of extrapolation. With the help of a computer program, statistical parameters are assigned to digitized weather patterns. These parameters include average intensity, variance, pattern bandedness, orientation of bands, ellipticity and characteristic pattern length. The cross-correlation coefficient γ is calculated efficiently by assuming it to be in a simple exponential form. The magnitude of its maximum value γ_{max} is an indication of pattern development, while the location of γ_{max} represents the best estimate of the pattern average motion. Wilson (1966) extended their work by examining the behaviour of γ_{max} as the time interval between patterns is increased, and he related the behaviour to precipitation type. He also investigated the predictability of the different wavelengths in an echo pattern and tentatively concluded that features of less than

20 km are "redundant and highly perishable information" which cannot be usefully forecast.

Actual forecasts and verifications of radar echo displacement have been attempted using several techniques. The first of these extrapolates individual echoes using a linear least-squares-fit through successive positions of their centroids. This "centroid" approach is reported by Wilk and Gray (1970) and Zittel (1976). This technique is successful for weather situations characterized by scattered isolated showers which are sufficiently compact to be defined by means of an appropriate threshold. Difficulties arise when echoes pass into or out of the field of view, merge, split or undergo drastic changes during successive time intervals. The second method is to combine the echoes into groups with similar characteristics which are expected to move in a common manner. This "cluster" approach, which requires a higher degree of computer sophistication, has been devised and tested with some success by Duda and Blackmer (1972) and updated by Blackmer *et al.* (1973). This technique, also called the SRI (Stanford Research Institute) model, has also been applied to track cloud motions by Wolf *et al.* (1977) and Endlich and Wolf (1981). It is conceptually appealing but may not be able to cope with complex weather systems without human assistance and is not suitable for rapid execution on a mini or microprocessor. A simpler technique, which we have adopted at McGill University, involves a "cross-correlation" technique applied to either the whole area of view of a single radar or to subareas of such a region. This technique has been described in Austin and Bellon (1974) and verified in a substantial real-time test (Bellon and Austin (1978)). This procedure is more appropriate for weather patterns with a relatively large characteristic scale length. Leese *et al.* (1971) and Endlich *et al.* (1971) also chose the cross-correlation method for obtaining cloud motions from satellite data. Muench (1976) experimented with a simple algorithm which follows the leading and trailing edge of weather patterns. Another scheme reported by Tatehira and Makino (1974) and Tatehira *et al.* (1976) bases the echo translation on the 700-mb wind field. This technique functions reasonably well in mid-latitudes where radar echoes frequently move with a velocity close to that of the wind at this level. In the tropics, however, there is little chance for the effective use of such a technique, since echo motion is less obviously related to the wind field at any height (Pestaina-Haynes and Austin, 1976).

The radar data used as input to these procedures must represent the required rain field as accurately as possible and be free of artifacts caused by ground clutter and beam blocking. At McGill we have chosen 2- or 3-km Constant Altitude PPI maps (CAPPI) while in the UK a pseudo CAPPI, made of four low-evelation angles, has been successfully used for the same purpose. Greene (1972) advocates the use of vertically integrated liquid water (VIL) maps, although CAPPI or at least pseudo CAPPI maps are required if quantitative surface rainfall forecasts are needed.

Elvander (1976) presented the results of an experiment where he compared the performance of the "centroid" method, the "cluster" model, and an earlier version of the McGill "cross-correlation" technique on the same digital PPI data set. The results were generally favourable to the McGill technique. It is particularly applicable to a single radar system and suitable for those parts of the world where the weather patterns display a motion in one predominant direction. A more complete summary of objective forecasting, using radar data, is provided by Collier (1978) who concludes by advocating a "man in the loop" with a least-square-fit technique for use in the UK system. He feels that the involvement of the man is the best way to quality-control the radar data and to perform the clustering function if required.

2 The McGill technique (SHARP)

Recently we have been working on a larger-scale version of the same type of "cross-correlation" technique using GOES satellite imagery which has been processed to yield areas of high probability of rain to interpolate between radars over an area of about (2000 x 1500) km^2. The earlier radar procedure is called SHARP (Short-term Automatic Radar Prediction) (Bellon and Austin, 1978) and the more recent large-scale system RAINSAT (Bellon *et al.*, 1980). The two systems will be described separately, but it should be appreciated that SHARP has been in operation for more than five years, whereas RAINSAT has become operational only in 1981. We now continue with a discussion of the SHARP procedure.

The crucial component in the McGill approach is the pattern recognition technique which finds the best match between the current map and the one stored from the previous hour. This is done by calculating the cross-correlation coefficient for all possible displacements until its maximum value γ_{max} has been included and identified. It is economically unsound, however, to attempt the calculation of γ for all lags in both Cartesian directions until γ_{max} has been found. Since an analysis of a simulated test of archived summer data showed a predominant motion from about 260°, a velocity of 30 km h^{-1} from that direction was chosen as the first estimate of pattern displacement. Other first guess procedures could also be adopted; for example, the 700-mb wind could also be used as in Tatehira *et al.* (1976). In subsequent forecasts, the computed velocity from the previous hour is used as the initial guess. The SHARP program has a number of features including routines designed to eliminate ground echo, to detect anomalous propagation, to allow for adverse effects on boundaries and to determine if there is sufficient areal coverage (~ 2%) within a range of 200 km to warrant the making of a forecast. Significant work has been done in successive approximation techniques to make this program run as quickly as possible. The employed version takes less than 60 seconds. The forecast is then made by translating the current map by the appropriate vector and removing the ground echoes. A geographical overlay is then applied.

Point forecasts are made for prescribed stations by looking up the trajectory from the grid points around them by amounts given by the translation vector. In order to take into account possible cross-range errors, a line trajectory is expanded into a sector of 16°. The largest rainfall intensity within each arc at constant time lags from the points of interest is registered. Analysis of historical data had demonstrated that a sector of this magnitude would include 60% of expected deviations from straight line motion. On request, additional products, such as maps of echo maximum height and displays of accumulated rain are also sent to the Forecast Office.

An additional pattern matching routine is employed to recognize the presence of anomalous propagation and automatically to issue messages warning of its presence. Such a task was made possible by the fact that anomalous propagation is associated with characteristically shaped radar returns from mountains in the southeast sector. Under conditions of anomalous propagation, ground echoes assume radically different aspects which are of greater intensity and extent than those seen under normal atmospheric conditions. We have found that this configuration can be recognized by an objective technique using the computer, even in the presence of some weather echoes (Bellon and Austin, 1977). Anomalous propagation was correctly identified 152 times out of a total of 1036 hours in the first year of operation. No forecasts were issued on these occasions. Since we encountered only 9 erroneous detections and failed to detect it 7 times, we are convinced of the utility of this technique. In the UK, where the radar can see over large plains, the areas of anoprop can be rather variable. However, we believe that anomalous propagation can to some extent be objectively detected and removed by the measurement of the vertical profile of reflectivity as described in Collier *et al.* (1980).

3 The verification procedure

It is difficult to achieve an objective verification which generates scores corresponding to a subjective assessment of the quality of the product. The fact that the operational forecasters generally liked the product is probably the best type of evaluation an operational procedure could obtain. Nevertheless, in order to permit testing of improvements, as when differential motion of echo patches, and development and dissipation of rain areas are incorporated, we have verified the procedure in a quantitative fashion. The verification analysis involved the comparison between forecast and actual displacement as well as a grid-area by grid-area comparison of the radar rainfall intensities. Results from the 1976 and

1977 real time test of 735 one-hour forecast displacements based on 60-minute history showed a spatial error of 26%. This error, defined as the per cent ratio of the difference between the forecast and actual displacement, is also independent of the forecast period for up to 5 hours. Since 1979, the forecast displacements are based on 30-minute history resulting in a significant reduction of the error from 26% to 16%.

Since the quantity of ground measurements is very small in Canada and can produce only a partial verification, each grid area of the radar matrix was considered as a verifying location using radar data alone. The comparison between forecast and actual precipitation can be accomplished only by ignoring the well-known discrepancies that can occur between the precipitation intensity at radar beam height and the ground (see Browning and Collier, this volume, pp. 47–61, as well as any inexactness resulting from the estimation of rainfall rate from the $Z = 200R^{1.6}$ relationship used.

Before undertaking any verification procedure solely from radar data, it is necessary to determine the range beyond which the data are unreliable and the regions where ground echo, or the shadow of nearby obstacles, renders the data unreliable. This can be achieved by summing the intensity levels at each grid point of all 3-km CAPPIs during the summer tests. All verification and data analysis presented here was restricted to a problem-free domain of radius approximately 200 km over which the total rainfall for the summer was essentially constant.

The parameter commonly used for verification purposes is the Critical Success Index (CSI) (Donaldson *et al.*, 1975), defined as

$$CSI = \frac{X}{X + Y + Z}$$

where X is the number of hits, Y the misses and Z the false alarms.

		Observed: Value less than threshold	Observed: Value equal to or greater than threshold
Forecast	Value less than threshold	W	Y
Forecast	Value equal to or greater than threshold	Z	X

In order to allow spatial errors of the order of 1 or 2 grid lengths (i.e. 6 or 12 km) the rainfall intensity within the four touching grid areas which is closest in magnitude to the forecast level is identified as the matched value. This

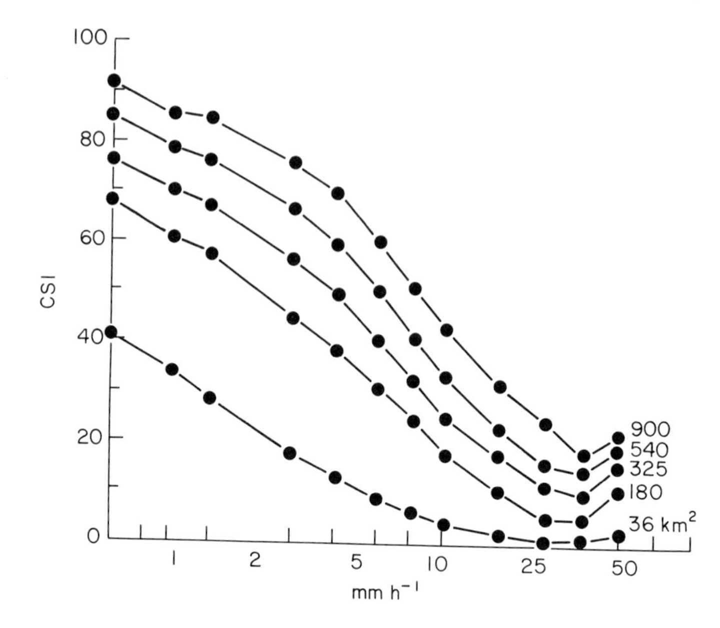

Fig. 1. CSI (Critical Success Index) as a function of rainfall rate with the verifying areas as a parameter based on 424 one-hour forecasts in 1976.

concept has been expanded to incorporate larger verifying areas consisting of (3 x 3), (3 x 5) and (5 x 5) matrices centred on the forecast grid area. Since the chance of exact matching is increased from 1 to 5, 9, 15 and 25, respectively, much improved numerical values are to be expected.

The dependence of CSI on the rainfall and size of verifying area is illustrated in Fig. 1. It demonstrates that the CSI scores decrease linearly with the logarithm of the rainfall rate, particularly for verifying areas other than the basic unit of 36 km^2. The general leveling of the scores for grid areas of 36 km^2 beyond 10 mm h^{-1} probably reflects the fact that the typical size of the rainfall areas of these intensities is down to one grid area.

The low CSI scores at high rainfall rates are influenced by their small probability of occurrence and relatively low persistence as well as their small scales. Thus, if a high rainfall rate can occur at a low probability, the chances of accurately forecasting it are dictated to a greater extent by its lifetime or probability of recurrence than by the decrease in accuracy of its forecast motion.

The decrease in accuracy verified over areas of 180 km^2 as the forecasts are extended beyond one hour is examined in Fig. 2. Here, the CSI scores for the various rainfall intensities form a family of curves plotted versus the logarithm of the forecast lead time. It reveals that the CSI scores decrease approximately exponentially with the length of the forecast for periods of up to 3 hours.

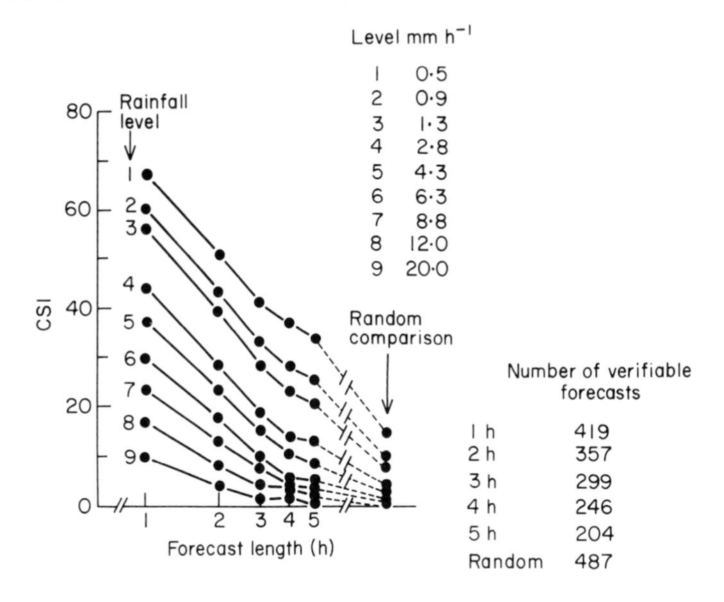

Fig. 2. CSI scores of various rainfall rate levels as a function of the logarithm of the forecast length (lead time) in hours verified over areas of 180 km^2 (1976 data).

Beyond the three-hour period, a reduction in the rate of decrease is experienced for all levels, but particularly for levels $\geqslant 4$, where the slope of the curves is nearly flat between the 4- and 5-hour interval. This behaviour implies that, under the steady-state assumption, the limit of the forecast skill for rates greater than 3 mm h^{-1} as verified over areas of 180 km^2, does not exceed 3 or 4 hours.

It is interesting to derive the lowest limit of CSI scores which indicates the total absence of a forecast skill. This is obtained by comparing two maps separated by such a large time interval (of the order of days) as to render them essentially independent, or random. The CSI scores for this type of comparison are presented at the right side of the plot in Fig. 2. The relatively small differences between the random scores and those of forecasts beyond 4 hours point to the limiting skill of the extrapolation forecast procedure.

While a CSI score provides an assessment of the expected quality of a forecast, its usefulness lies mainly in its role as comparator between different forecasting techniques, or as a barometer for evaluating various alternatives within one general technique. Comparisons have been derived for the forecasts made during the real-time test with what would have been obtained if predicted vectors were equal to the actual motion during the verifying interval; we label this as "optimum" in Fig. 3. Two additional techniques were evaluated: the assumption of no motion of the weather patterns and that of constant motion for the entire

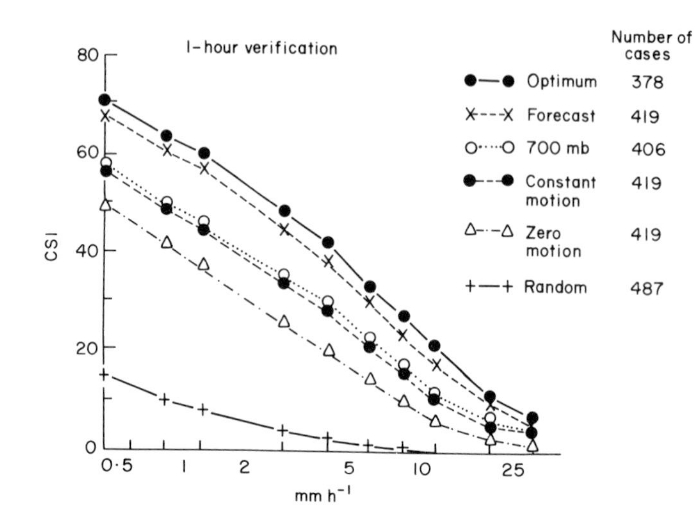

Fig. 3. CSI scores as a function of rainfall rate for various predictors verified over areas of 180 km^2.

summer. For the latter, 40 km h^{-1} from 260° was chosen, which corresponds to the average velocity for 1976 data. The results for verifying areas of 180 km^2 are representative of the other areas and are shown in Fig. 3. An immediate conclusion which can be drawn is that the score would have been only marginally increased if the overall motion of the weather systems had been perfectly forecast. This implies that a major portion of forecastability loss is attributed to the rearrangement and restructuring of weather patterns as they underwent development or decay during the forecast period, a direct consequence of the "steady state" assumption. The incorporation of a successful forecasting technique encompassing differential motion and echo growth and decay could in principle improve the accuracy which can be obtained. Results in Canada indicate that the advantages gained by these procedures are minimal for the scale of single radar coverage (Tsonis and Austin, 1981). But for a radar network (or satellite coverage) differential motion is important as discussed below.

4 Geographical distribution of growth and decay

The test described in Bellon and Austin (1978) provided an ideal data base for the determination of the geographical distribution of development and dissipation of summer storm systems in Eastern Canada. The areas of preferential growth and decay were identified by comparing forecast rain maps with observed radar rain patterns. In general it is found that, for these summer convective cases, the orographic effects are weak and ill defined.

It is the belief of the authors that the assimilation of preferred behaviour of rain patterns which enables a forecasting technique to depart from the "steady state" assumption, based on a statistical map of average orographic effect, would improve the accuracy of the forecasts only slightly in the Montreal region. This conclusion is supported by the observation that the growth areas identified in data from two sequential, largely convective summers showed significantly different patterns. However, in other parts of the world, the enhancement of a precipitation system can be significant and the neglect of orography will lead to a misleading product, especially in frontal situations (e.g. South Wales in Britain, as described by Browning *et al.*, 1975). This orographic effect can be quantitatively estimated as described by Browning and Hill (1981).

5 Satellite-based systems for forecasting precipitation over larger areas (RAINSAT)

An algorithm capable of inferring rain areas from IR and VIS satellite imagery was devised by Lovejoy and Austin (1979). It therefore seemed sensible to combine the latter technique with that already tested for forecasting the movement of radar echoes. The increased coverage includes the area of Canada and the adjacent parts of the United States viewed by SMS/GOES-E, totalling $(1500 \times 2000)\,\text{km}^2$. This area would require a network of 20 radars to give adequate coverage.

The data described here consist of SMS/GOES-E images archived at 30-minute intervals initially obtained from the University of Wisconsin and more recently from the Satellite Data Laboratory of the Atmospheric Environment Service. Corresponding radar data were obtained from the McGill Radar Weather Observatory located 20 km west of Montreal. These were in the form of CAPPI maps for 3-km altitude extending out to a range of 200 km and digitized in levels providing the average rainfall rate over $(4 \times 4)\,\text{km}^2$ grid areas.

Because of the inflexibility of some satellite display systems, one is often obliged to remap other data sets, e.g. radar data, onto satellite coordinates. Since in this work a pattern recognition technique is applied to two successive satellite pictures, it is convenient to use constant grid maps showing minimal areal and directional distortion. Moreover, since comparisons will ultimately be performed between the satellite-derived rain maps and synoptic charts, whether as part of a research analysis, or in real time by a forecaster on duty, it is highly desirable to produce products on equivalent map projections. In mid-latitudes, a conic projection of the earth, true at two standard parallels, has been commonly used by meteorologists. Thus, using subroutines of the navigational model described

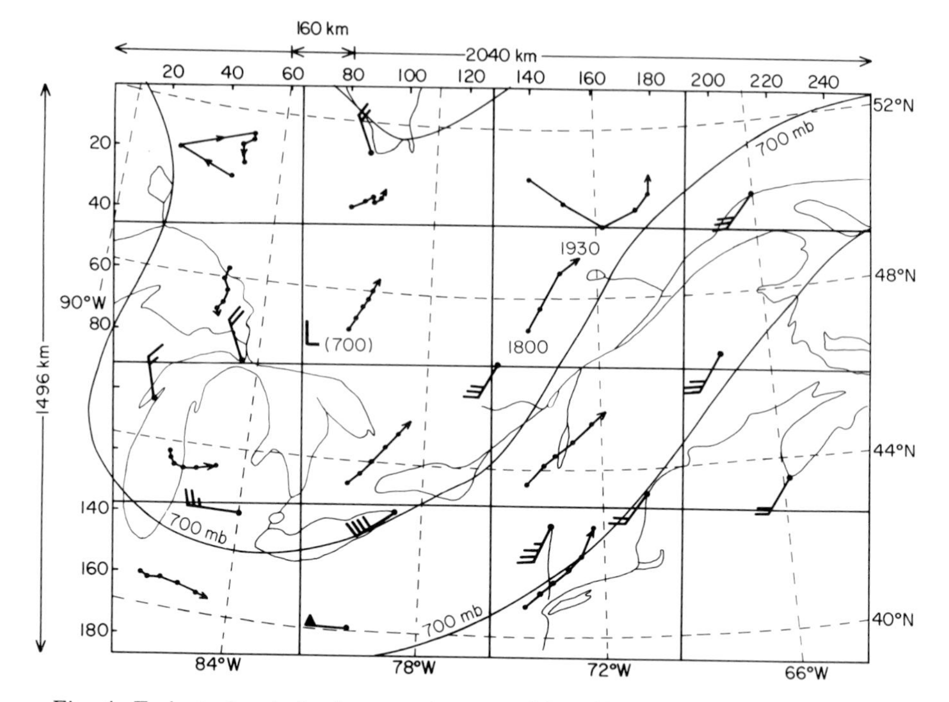

Fig. 4. Trajectories derived over sub-areas with sufficient satellite-derived rain area coverage for 1 June 1977. Arrows represent rain area trajectories; barbs, the 700-mb wind field (full barb = $5\ \mathrm{ms}^{-1}$) and the lines the 700-mb analysis. See text for further details.

by Smith and Phillips (1972), the archived satellite data have been remapped on a (187 x 255) matrix with a constant grid resolution of 8 km. Maps of raining areas are then produced by analysis of the visible and IR data and verified by the data from a second radar located in Toronto, approximately 450 km south-west of Montreal.

In order that satellite rain maps may be used as input to a forecasting scheme, the technique must produce maps of the zones which are sufficiently consistent in time to be trackable by a pattern recognition algorithm. It is also apparent that this algorithm must be applied to sub-areas of the entire map. Hence, the (187 x 255) matrix has been sectioned into 16 sub-areas, and thus this method behaves in a similar way to the "cluster" technique described earlier. This gives a (46 x 64) array, or $(368 \times 512)\ \mathrm{km}^2$. Only zones with a minimum of 1% coverage of rain are considered in this system. The 700-mb wind field could probably be used for similar purposes, but is not generally available with high spatial and temporal resolution.

Typical results of the tracking program are illustrated in Fig. 4. More detailed analyses of several cases are given in Bellon *et al.* (1980). The half-hourly displacements, computed in fractions of grid lengths to avoid quantization effects, have been summed to form a trajectory for each sub-area. Unless otherwise indicated, each trajectory in Fig. 4 begins at 1730 GMT and terminates at 2000 GMT. The 700-mb analysis for 2400 GMT, with the radiosonde wind data, has been added for visual comparison. Figure 4 shows that the rain zones on the southern section of the map have been correctly followed, yielding a displacement which compares favourably with the 700-mb wind. Speeds of the order of 50 and 60 km h^{-1} correlate well with 700-mb wind speeds 15 to 20 m s^{-1} found ahead of the 700-mb trough. In this region, the linearity in both speed and direction augurs well for a very-short-range forecasting system based on pattern extrapolation. Behind the arctic front, or closer to the 700-mb trough, lower speeds are calculated in accordance with upper level data, but, as exemplified by sub-area (3, 1) which includes the Michigan area, the expected passage of an upper front must be taken into account when performing a forecast by pattern advection. A change in speed and direction can be inferred from the monitoring of satellite data only after its occurrence. Advance information may be provided by prognostic models operating in the 6- to 12-hour period, including possible development or dissipation of rain zones.

A disturbing aspect of Fig. 4 is the trajectories in the northern portion of the map. It is obvious that the first two calculated displacements for sub-areas (1, 1) and (1, 3) constitute spurious motion. These are caused by the absence of easily recognizable features, or possibly by the creation of artificial ones arising from the arbitrary sectioning of patterns into sub-areas. Experience has demonstated that a coverage greater than 40% is as detrimental to a pattern recognition technique as insufficient coverage, by underestimating the actual motion and hence being subjected to quantization effects. An additional explanation for the erratic motion in sub-areas (1, 1) and (1, 3) is that the rain determination algorithm, derived from radar data near 45°N latitude, may not be as stable in northern latitudes, with significantly different climatology. However, as was the case in the real-time operation of SHARP, criteria are being developed which can reject apparently unacceptable displacements.

6 Conclusions

In the Montreal area, we have operated a real-time very-short-range forecasting system based upon a single radar continuously since 1976. The technique uses simple extrapolation with a single translation vector obtained from a cross-correlation algorithm applied to the entire area of interest. The radar data used are 2- or 3-km Constant Altitude PPI maps from which most of the artifacts have been objectively removed. The system has been well received by operational personnel who have found the products useful for a variety of their forecasting tasks. These include very-short-range forecasting of the onset of rain, estimation of forest fire indices and severe weather warning for airports and for the general public.

The analysis of results in terms of Critical Success Index indicated that the forecast technique adopted represents nearly the best accuracy that can be attained under the "steady-state" assumption. The major portion of forecastability loss is attributed to the development and dissipation of weather patterns rather than to errors in their forecast motion. In fact, attempts to improve the accuracy of the system by incorporating second-order effects in echo growth and decay, orographic effects and differential motion have been found to give very little improvement.

A system covering a larger area, employing combined radar systems and geostationary satellite data, has been put into real-time operation more recently. Initial results using a 16-sub-area cross-correlation tracking algorithm have been quite encouraging.

In general, we have found that the linearly extrapolated very-short-range forecast products are useful in the 0- to 3-hour domain. It remains to be seen whether any of the more sophisticated developments of such systems, or even real-time incorporation of other data, will significantly improve the forecast accuracy. The results reported here are for the Montreal area but are probably applicable in other mid-latitude regions.

Acknowledgments

Much of the original research work described here was carried out with Dr Shaun Lovejoy. Assistance in computing was provided by Dr S. Radhakant and Ms Alamelu Kilambi. The projects have been supported by the Satellite Data Laboratory of the Atmospheric Environment Service and the Natural Sciences and Engineering Research Council of Canada.

References

Austin, G. L. and Bellon, A. (1974). The use of digital weather records for short-term precipitation forecasting. *Q. J. R. met. Soc.* **100**, 658–664.

Bellon, A. and Austin, G. L. (1977) Short-term automated radar prediction. Final report for AES (DSS) Contract OISU.KM 601–4–758. (Available from the Stormy Weather Group, McGill University, Montreal, PQ, Canada).

Bellon, A. and Austin, G. L. (1978). The evaluation of two years of real time operation of a short-term precipitation forecasting procedure (SHARP), *J. appl. Met.* **17**, 1778–1787.

Bellon, A., Lovejoy, S. and Austin, G. L. (1980). Combining satellite and radar data for the short-range forecasting of precipitation. *Mon. Weath. Rev.* **108**, 1554–1566.

Blackmer, R. H., Duda, R. O. and Reboh, R. (1973). Application of pattern recognition to digitised weather radar data. Final report Contract 1–36072, SRI Project 1287, Stanford Research Institute, Menlo Park, California, pp. 89.

Boucher, R. J. (1961). The motion and predictability of precipitation areas as determined from radar observations. Proceedings, 9th Weather Radar Conference, pp. 37–42. American Meteorological Society, Boston, Massachusetts.

Boucher, R. J. (1963). Radar precipitation echo motion and suggested prediction techniques. Proceedings, 10th Weather Radar Conference, pp. 1–7, American Meteorological Society, Boston, Massachusetts.

Browning, K. A. and Hill, F. F. (1981). Orographic rain. *Weather* **36**, 326–329.

Browning, K. A., Pardoe, C. W. and Hill, F. F. (1975). The nature of orographic rain at wintertime cold fronts. *Q. J. R. met. Soc.* **101**, 333–352.

Collier, C. G. (1978). Objective forecasting using radar data: a review. Research Report No. 9, pp. 16. Meteorological Office, Radar Research Laboratory. RSRE Malvern, England.

Collier, C. G., Lovejoy, S. and Austin, G. L. (1980). Analysis of bright bands from 3-D radar data. Proceedings 19th Weather Radar Conference, pp. 44–47. American Meteorological Society, Boston, Massachusetts.

Donaldson, R. J., Dyer, R. M., and Kraus, M. J. (1975). An objective evaluation of techniques for predicting severe weather events. Preprints, 9th Conference on Severe Local Storms (Norman) pp. 321–326. American Meteorological Society, Boston, Massachusetts.

Duda, R. O. and Blackmer, R. H. (1972). Application of pattern recognition techniques to digitised weather radar data. Final Report covering the period 25 May 1971 to 31 March 1972, Contract 1–36072, SRI Project 1287, Stanford Research Institute, Menlo Park, California, pp. 135.

Elvander, R. C. (1976). An evaluation of the relative performance of three weather radar echo forecasting techniques. Preprints, 17th Radar Meteorology Conference (Seattle), pp. 526–532. American Meteorological Society, Boston, Massachusetts.

Endlich, R. M. and Wolf, D. E. (1981). Automatic cloud tracking applied to GOES and METEOSAT observations. *J. appl. Met.* **20**, 309–319.

Endlich, R. M., Wolf, D. E., Hall, D. J. and Brain, A. E. (1971). Use of a pattern recognition technique for determining cloud motions from sequences of satellite photographs. *J. appl. Met.* **10**, 105–117.

Greene, D. R. (1972). A comparison of echo predictability: constant elevation vs VIL radar-data patterns. Preprints, 15th Radar Meteorology Conference, pp. 111–116. American Meteorological Society, Boston, Massachusetts.

Hilst, G. R. and Russo, J. A. (1960). An objective extrapolation technique for semi-conservative fields with an application to radar patterns. Technical Memo 3, The Travelers Weather Research Center, Hartford, Connecticut.

Kessler, E. (1961). Appraisal of the use of radar in observation and forecasting. Proceedings 9th Weather Radar Conference, pp. 13–36. American Meteorological Society, Boston, Massachusetts.

Kessler, E. (1966). Computer program for calculating average lengths of weather radar echoes and pattern bandedness. *J. atmos. Sci.* **23**, 569–574.

Kessler, E. and Russo, J. A. (1963). Statistical properties of weather radar echoes. Proceedings, 10th Weather Radar Conference (Washington, DC), pp. 25–33. American Meteorological Society, Boston, Massachusetts.

Leese, J. A., Novak, C. S. and Clark, B. B. (1971). An automated technique for obtaining cloud motion from geosynchronous satellite data using cross-correlation. *J. appl. Met.* **10**, 118–132.

Ligda, M. G. (1953). The horizontal motion of small precipitation areas as observed by radar. Technical Report 21, Department of Meteorology, M.I.T., Cambridge, Massachusetts, pp. 60.

Lovejoy, S. and Austin, G. L. (1979). The delineation of rain areas from visible and IR satellite data for GATE and mid-latitudes. *Atmosphere Ocean* **17**, 77–92.

Muench, H. S. (1976). Use of digital radar data in severe weather forecasting. *Bull. Am. met. Soc.* **57**, 298–303.

Noel, T. M. and Fleisher, A. (1960). The linear predictability of weather radar signals. Research Report 34, Department of Meteorology, M.I.T., Cambridge, Massachusetts, pp. 46.

Pestaina-Haynes, M. and Austin, G. L. (1976). Comparison between Maritime Tropical and continental mid-latitude precipitation lines. *J. appl. Met.* **15**, 1077–1082.

Russo, J. A. and Bowne, N. E. (1961). Linear extrapolation as a meteorological forecast tool when applied to radar and cloud ceiling patterns. Proceedings, 9th Weather Radar Conference, pp. 43–49. American Meteorological Society, Boston, Massachusetts.

Smith, E. A. and Phillips, D. R. (1972). Automated cloud tracking using precisely aligned digital ATS pictures. IEEE Trans., Comput. C-21, pp. 715–729.

Tatehira, R. and Makino, Y. (1974). Use of digitised echo pattern for rainfall forecasting. *J. met. Res., Japan met. Agency* **26**, 188–199.

Tatehira, R., Sato, H. and Makino, Y. (1976). Short-term forecasting of digitised echo pattern. *J. met. Res., Japan met. Agency* **28**, 61–70.

Tsonis, A. A. and Austin, G. L. (1981). An evaluation of extrapolation techniques for the short-term prediction of rain amounts. *Atmosphere Ocean,* **19**, 54–56.

Wilk, K. E. and Gray, K. C. (1970). Processing and analysis techniques used with the NSSL weather radar system. Preprints, 14th Radar Meteorology Conference, pp. 369–374. American Meteorological Society, Boston, Massachusetts.

Wilson, J. W. (1966). Movement and predictability of radar echoes. US Weather Bureau, Contract CWB–11093, The Travelers Weather Research Center, Hartford, Connecticut.

Wolf, D. E., Hall, D. J. and Endlich, R. M. (1977). Experiments in automatic cloud tracking using SMS-GOES data. *J. appl. Met.* **16**, 1219–1230.

Zittel, W. D. (1976). Computer applications and techniques for storm tracking and warning. Preprints, 17th Radar Meteorology Conference (Seattle), pp. 514–521. American Meteorological Society, Boston, Massachusetts.

3.4

Use of a Conceptual Model of the Life-cycle of Mesoscale Convective Systems to Improve Very-short-range Forecasts

EDWARD J. ZIPSER

1 Introduction

The purpose of this chapter is to indicate ways that conceptual models of the life-cycle of mesoscale convective systems can lead to improvements in forecasting compared with what is achieveable by simple extrapolation. While mesoscale convective systems are used as an example of this approach, their importance for the very-short-range forecast problem should not be underestimated. In the tropics, and in mid-latitudes in the warm season, a large fraction of the total precipitation at any location falls from atmospheric structures organized on the mesoscale, with horizontal dimensions up to a few hundred kilometres (Houze, 1981). Awareness of the typical evolution of such systems can improve interpretation of a real-time mesoscale data stream.

Simply put, when systems change little during the forecast period, simple extrapolation models (Austin and Bellon, 1982) will be useful. The approach described here will be most useful when change in the system could be expected. For very-short-range forecasts of one to ten hours, it is of importance to anticipate changes in mesoscale convective systems.

For current purposes, mesoscale convective systems are defined as cloud and precipitation systems, together with their associated circulation systems, which include a group of cumulonimbus clouds during most of the lifetime of the system. The cumulonimbus cloud group must exist for several lifetimes of its constituent clouds (say for at least two hours), and the cumulonimbus group must contribute at some time to a common upper tropospheric shield of outflow air. Its convective-scale downdraughts must also merge at some time to form a continuous zone of cool air in the low troposphere. Normally, extensive stratiform precipitation would fall from the outflow shield, evaporating to a greater or lesser extent before reaching the ground. This and other structural features will be described as functions of the life-cycle.

2 Structure of mature systems

One system clearly satisfying the above definition is the tropical squall line. An early descriptive study by Hamilton and Archbold (1945) was followed by increasingly quantitative papers based upon the results of recent tropical field programs (e.g. Zipser, 1969, 1977; Houze, 1977; Gamache and Houze, 1982). The structure of mature, steady-state systems, to the extent that they exist, has been well documented. The mature tropical squall line is characterized by convective cells along the leading edge, moving rapidly, with an extensive trailing "anvil" region, often precipitating over 100 km or more in width, with mesoscale ascent helping to maintain the stratiform upper clouds, and mesoscale descent in the low troposphere in the stratiform rain area.

The Global Atmospheric Research Program's Atlantic Tropical Experiment (GATE), carried out in the tropical eastern Atlantic in 1974, resulted in many case studies of mesoscale phenomena. For the GATE area, it has been established that most mesoscale precipitation features are similar in structure (Leary and Houze, 1979; Houze and Betts, 1981). This is true for both squall and "non-squall" systems, which are distinguished from each other by the faster forward motion and stronger surface winds of the squall line systems. Figure 1 illustrates the Leary and Houze model of a generalized mesoscale precipitation feature. Typical airflow circulations with respect to this type of radar structure are illustrated in Zipser (1977), Houze (1977), and Houze and Betts (1981).

The reader might well ask why we should compare these relatively modest tropical systems to those of mid-latitudes, which are known to be associated with severe weather such as hail and tornadoes. The fact is that for understandable reasons, most studies of mid-latitude mesoscale systems have emphasized severe weather almost to the exclusion of the generic nature of the entire phenomenon. It is important to distinguish between *organized* convection and *severe* convection, as there are many examples of one without the other. In the United States, many anecdotes could be related about researchers focusing radars on severe convective cells at the leading edge of a squall line, ignoring massive stratiform precipitation, even driving home through it, and later questioning its existence.

With the notable exception of the supercell storm (Browning, 1977), there is little doubt that in their mesoscale aspects, mid-latitude and tropical squall lines have many features in common. An early analysis of a UK squall line (Pedgley,

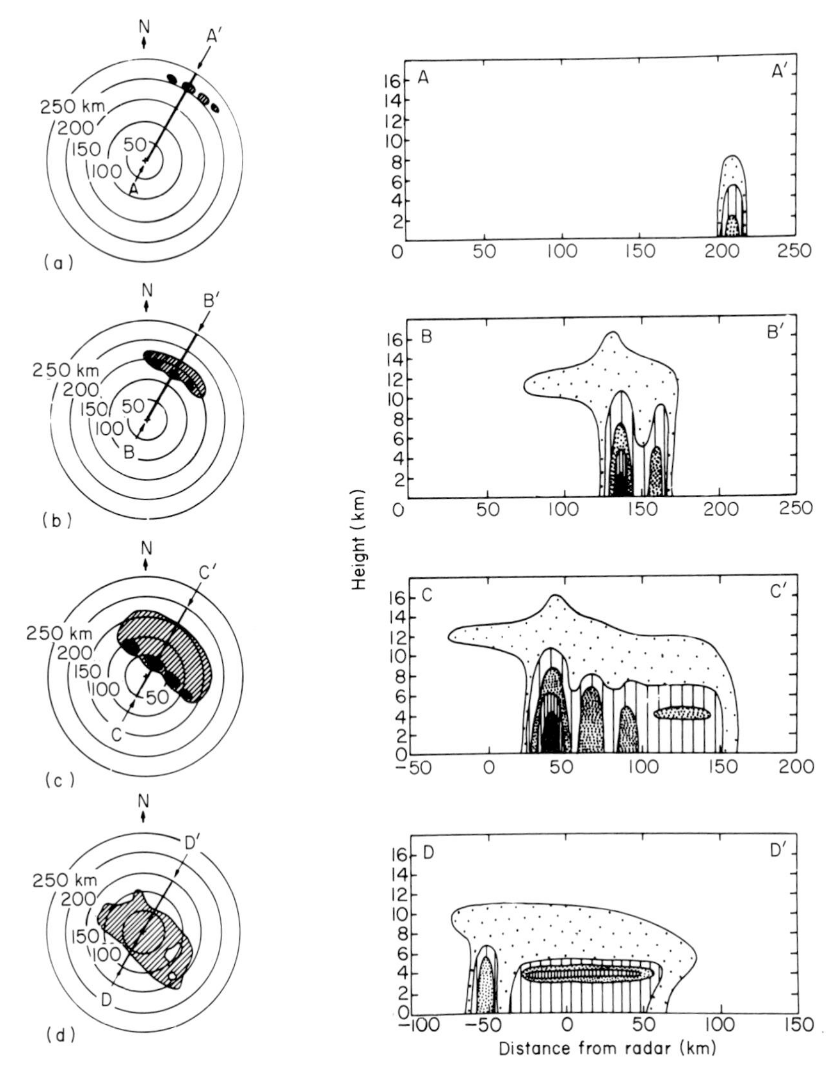

Fig. 1. Schematic of the structure of a mesoscale convective system as viewed by radar in horizontal and vertical cross-sections during (a) formative, (b) intensifying, (c) mature, and (d) decaying stages of its life-cycle. (After Leary and Houze, 1979; see Houze, 1977, Zipser, 1977, and Zipser *et al.,* 1981, for examples of wind flow relative to such systems.)

1962; Fig. 2) shows a trailing precipitation region from the "anvil" which is remarkably similar to the tropical cases. Newton (1963), Sanders and Emanuel (1977), and Ogura and Liou (1980) performed analyses of squall lines in the

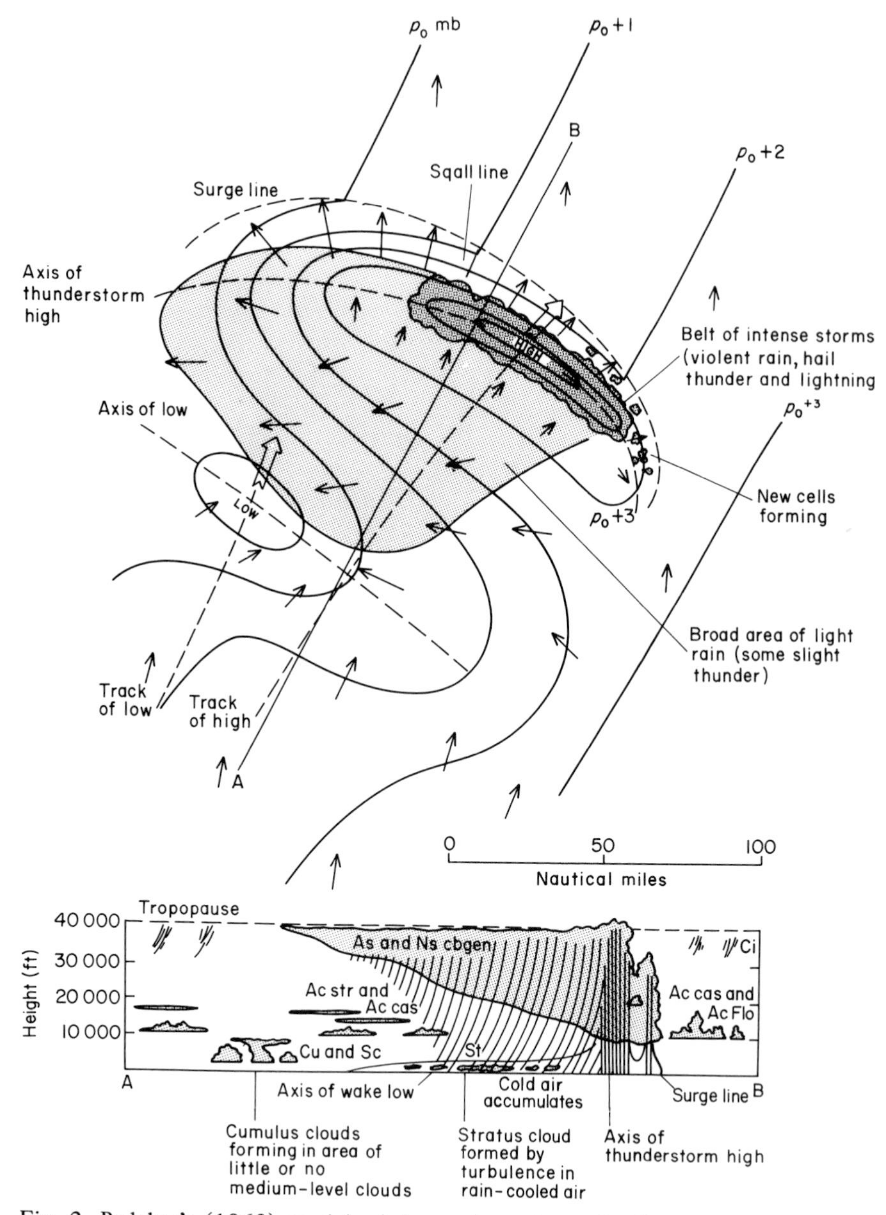

Fig. 2. Pedgley's (1962) model of the mature stage of a "meso-scale thunderstorm system" over the UK as the result of a case study. Areas of light and heavy precipitation are indicated; arrows show surface wind direction. The schematic vertical cross-section from A to B is the bottom panel. The system motion is from A toward B.

central United States in which the extensive trailing precipitation and airflow pattern normal to the system was in most respects similar to tropical squall lines.

Recently in the United States, increasing attention has been given to a class of system called "mesoscale convective complexes" by Maddox (1980). His outstanding contribution has been to show how frequent these systems are in the United States in the warm season, and that they contribute much of the precipitation as well as much of the severe weather. This is in spite of Maddox's particular definition of these complexes, which includes only the largest and most intense ones. While he shows mostly synoptic and satellite data, as research quality radar data is scarce, there is little doubt that their structure is also similar to that of the squall lines in many respects.

Figure 3, one of Maddox's (1980) examples of a mesoscale convective complex during its mature stage, illustrates the strong convective line along the leading edge of the system, and the extensive stratiform region to the north. At one station passed by the system, an initial burst of heavy rain was followed by more than $4\frac{1}{2}$ hours of continuous light rain and thunder. It is precisely the propensity for widespread thunder to be associated with the *stratiform* (!) precipitation which has led to confusion about just what kind of precipitation exists to the rear of the squall line or complexes. It is of interest that the UK system of Fig. 2 has thunder reported in the stratiform precipitation area, as do many tropical systems.

The topic of this book is very-short-range forecasting. To the extent that these mesoscale convective systems are in steady state, extrapolation may be successful. However, with the exception of systems which move very rapidly, so as to cover a great deal of territory during their quasi-steady-state phase, it is the usual experience that mesoscale convective systems change drastically over periods of several hours. Thus, we turn our attention to the way they evolve.

3 Life-cycle

There is no satisfactory theory of mesoscale convective systems to appeal to for an explanation of why and how they evolve. In all probability, a crucial factor is the inevitable production of precipitation-induced downdraughts, which in turn produce cold air at low levels and mesoscale high pressure at the surface, which in turn produces strong mesoscale wind anomalies. These can easily counteract any tendency for the system (if such exists) to seek equilibrium with the larger-scale forcing.

Consider for simplicity the following hypothetical example: a two-dimensional convergence zone in the lowest kilometre about 300 km wide and infinitely long (Fig. 4). In the first case (top) the horizontal mass convergence, and therefore the vertical mass flux through the 1-km level is six units, evenly

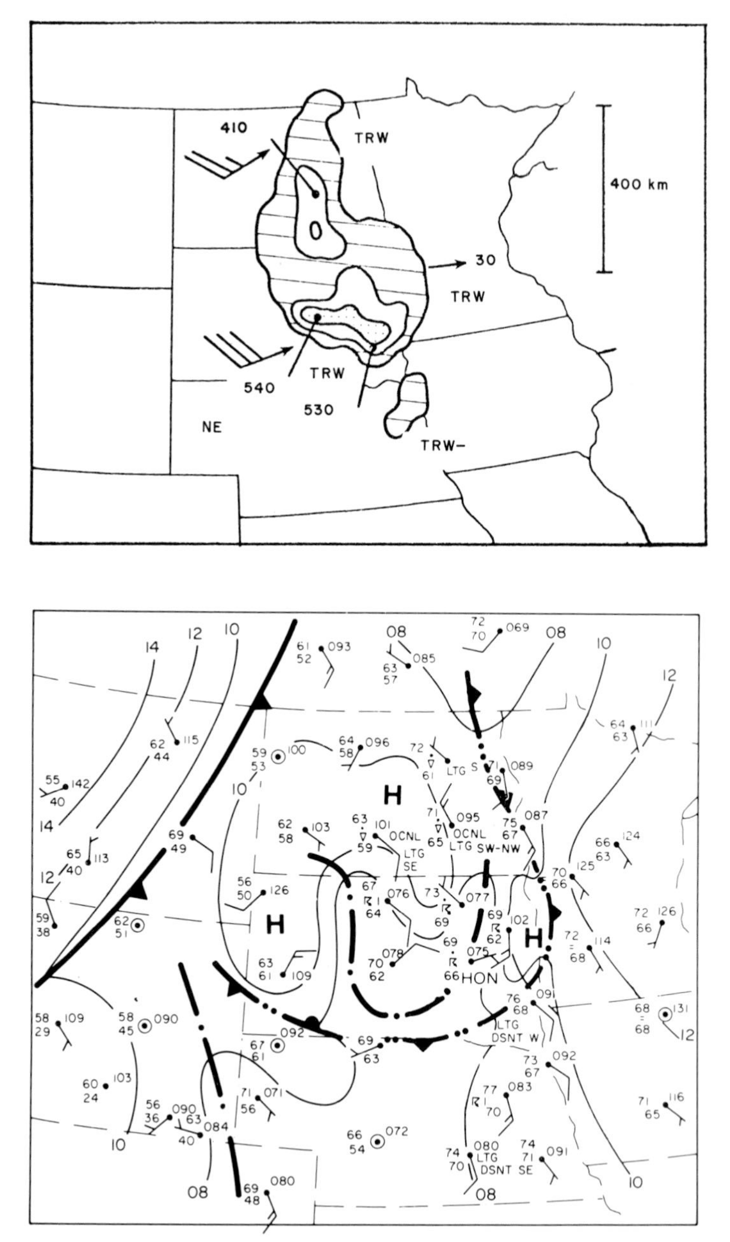

Fig. 3. Radar summary chart for 0835 GMT 12 July 1979 (top) and surface analysis for 0900 GMT 12 July 1979 (bottom). Radar contours are reflectivity levels, with maximum tops given in hundreds of feet. Surface winds are for full barb = 5 m s^{-1}. (After Maddox, 1980.)

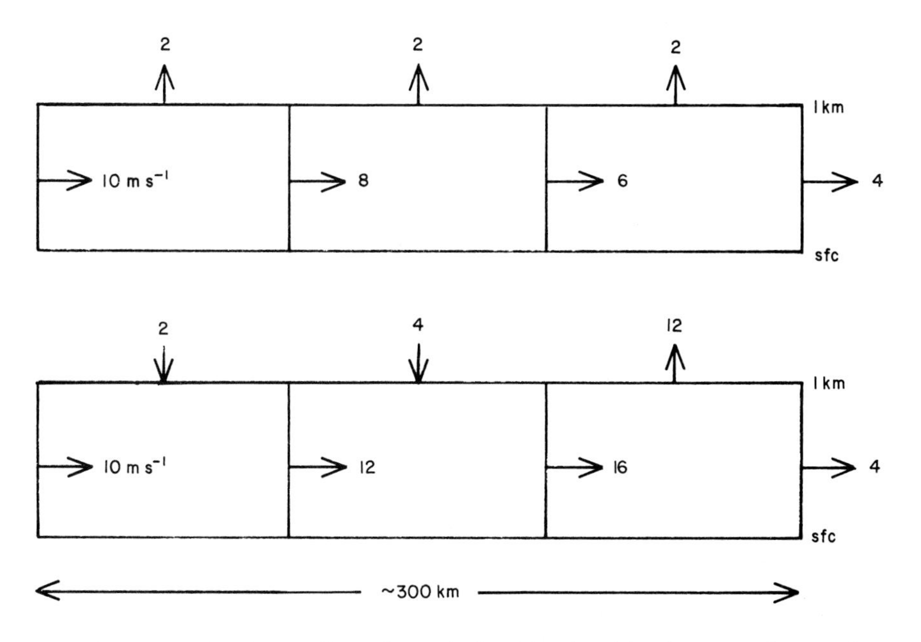

Fig. 4. Schematic of mass flow through the 1-km level in the case of synoptic-scale convergence distributed uniformly in the horizontal (above) and in the case of a squall-line type disturbance (below). Both cases represent a vertical plane across a two-dimensional system. The horizontal arrows represent wind speed and the vertical arrows mass flux. The units for mass flux are approximately $\times 10^3$ kg s^{-1} per metre of distance into the page. To compare with the last figure, for a system 200 km long, the units are approximately $\times$ (2×10^8 kg s^{-1}).

distributed. In the other case (bottom), which is representative of a squall line, the *net* mass flux is still six units, but the wind field within the zone is completely different. In fact, twelve units of low-level air are rising through the 1-km level, partially compensated by six units of sinking. In a real squall line the excess ascent often takes place entirely within a zone of 10–30 km width, stripping off the warm, moist boundary layer air and replacing it with boundary layer air which cannot support buoyant convection. If we assume that the large-scale convergence can supply six units of low-level air, but not twelve, such a system *cannot* remain coupled to the large-scale convergence, but must either dissipate quickly, or it must propagate into new air at a rapid rate. For example, the squall lines in West Africa and the adjacent Atlantic Ocean typically move at twice the speed of the synoptic-scale easterly waves which presumably had something to do with their generation.

Fritsch *et al.* (1976) have noted this imbalance between the synoptic-scale mass and moisture flux, or *large-scale supply rate*, and the *much larger mesoscale consumption rate*, and the importance of this fact for cumulus parameterization

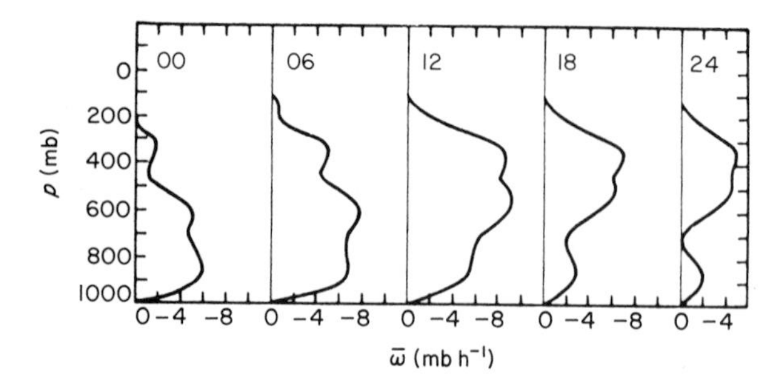

Fig. 5. Average vertical motion profiles obtained from observed winds over the GATE ship array at 6-h intervals on a particular day dominated by a few mesoscale convective systems whose rainfall peak was between 12 and 18-h (after Betts, 1978).

theories. It can be surmised that where this imbalance is greatest, i.e. where strong mesoscale systems form in the absence of strong synoptic-scale ascent, those mesoscale systems will monopolize the limited large-scale supply of boundary layer air, and few other systems are likely to exist nearby, in space or in time. Examples would be weak easterly waves triggering strong tropical squall line systems, or weak mid-latitude cold fronts triggering strong mesoscale convective complexes. On the other hand, a very strong spring cyclone in the midwestern United States has been known to contain several strong mesoscale systems which can also be long-lived.

Given that mesoscale convective systems evolve, how do they do so? The Leary and Houze model of the life-cycle of a mesoscale precipitation feature (Fig. 1) explains many of the observations in the tropical systems. The formation and expansion of the stratiform precipitation region is associated with the mesoscale downdraught, the absence of much cloud below the melting level, and the common presence of a melting band. It is becoming apparent that independent ascent is taking place aloft, to the extent that the stratiform precipitation is greater than could be accounted for by assuming passive particle descent, aggregation, etc., from the cumulonimbus debris in the divergent outflow aloft (Gamache and Houze, 1982). In that light, it is relevant to note that as systems mature, the vertical velocity maximum on the mesoscale, usually found lower during the formative stages, rises to the upper troposphere with time.

Examples of this typical time sequence are given in Figs. 5 and 6. Betts' (1978) computations are of mean vertical motion over the entire GATE ship array 700 km across, while Ogura and Chen's (1977) are for an Oklahoma sounding network the average spacing of which was 85 km. The difference in spacing accounts for most of the difference in magnitude of the vertical motion, as the GATE array is far too large for the vertical motions in Fig. 5 to be as

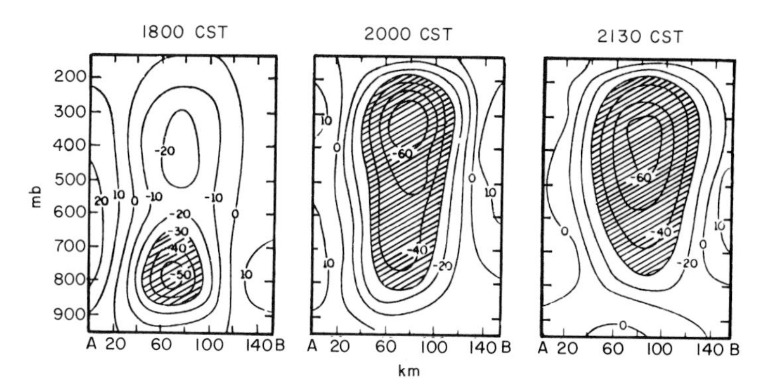

Fig. 6. Time sequence of the distribution of vertical motion (10^{-3} mb s^{-1}) in a vertical cross-section through an Oklahoma squall line, obtained from observed winds from a dense rawinsonde array (after Ogura and Chen, 1977, figure slightly abridged).

intense as they are in an individual mesosystem. Note that despite these differences, the vertical motion profiles evolve in the same way. In the Oklahoma case, the downward motion at low levels in the latest stage is explicit, while it may be surmised to exist in Fig. 5. Other GATE cases (Gamache and Houze, 1982) find it explicitly. The important point is that the mesoscale convective system evolves from an early intensifying stage in which intense convection is dominant, with a strong net upward motion at low levels, through the mature stage into a decaying stage in which convective rain exists, but becomes less important, while stratiform rain associated with upward motion at high levels becomes predominant.

The typical evolution of a system from mostly convective to mostly stratiform rain is illustrated in Fig. 7. This example is of a relatively small and weak system in GATE, but is a sequence believed to be representative of mesoscale convective systems in general. During a time period when the total area covered by precipitation, and the total rain volume, were not changing much with time, the way that the rain was partitioned into convective and stratiform character was evolving markedly. (Note that the rain volume scales are not the same.) The convective rain falls at heavy rates throughout, and the anvil-type rain falls at light to moderate rates throughout, but the relative amounts of area covered by the two regimes changes systematically, with the convective rain all but disappearing within an hour or two more. At that time, the anvil rain also begins to taper off and end, and mesoscale subsidence produces residual dry air at low levels as the entire episode ends.

This sort of time sequence is presented in schematic form in Fig. 8 for a prototype system going through the Leary and Houze life-cycle (Fig. 1), of size

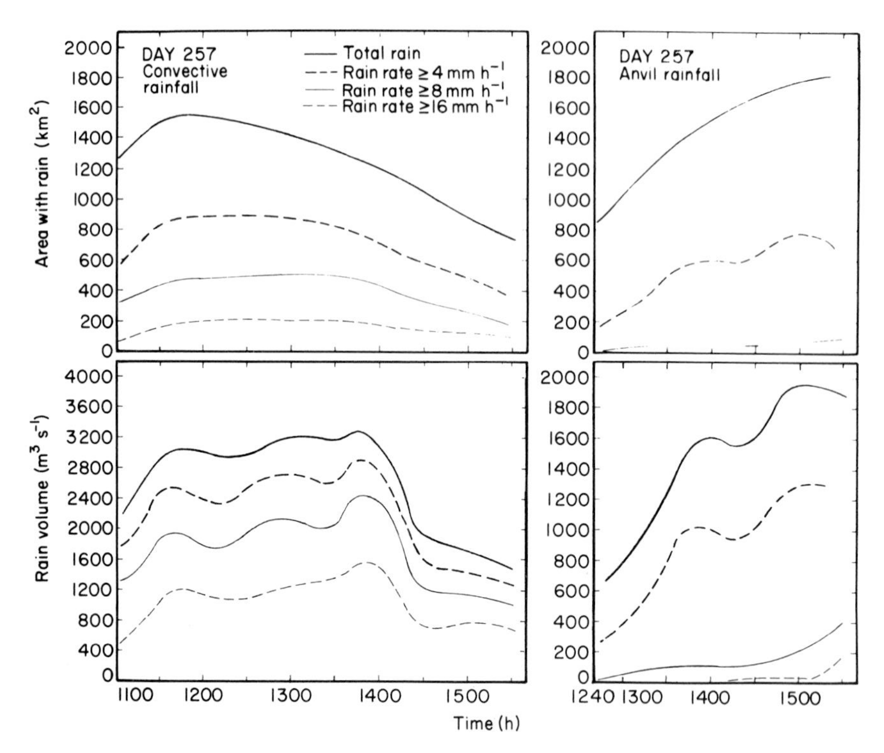

Fig. 7. For a quasi-two-dimensional portion of a GATE non-squall mesoscale system, time history of area covered by precipitation (top) and rain volume (bottom), above a threshold rain rate of 0.5 mm h^{-1} and also above threshold rain rates 4, 8, and 16 mm h^{-1}. The relevant portion of the convective band was 66 km long and averaged 30 km in width. The convective rainfall along the leading edge (left) is analysed separately from the trailing stratiform rain (called "anvil." right). (After Zipser *et al.,* 1981.)

and strength equivalent to a vigorous GATE system. This life-cycle model does not assume any period of near-steady-state conditions; if such were to exist one should envisage all curves becoming quasi-horizontal during the mature stage before resuming their evolution into the decay phase.

In the present schematic (Fig. 8), an initial convergence zone which could be mesoscale or small synoptic scale, manifested by an increase in M_{net}, is followed by the outbreak of organized convection within two hours (rapid rise in upward convective mass flux, M_{u}). With a small time lag, total rainfall rate (R) rises correspondingly; initially it is 100% convective. Also with a small time lag, a rapid rise in downdraught mass flux (M_{d}) follows; this is also initially 100% convective-scale. Over the next several hours, stratiform rain $(R)_{\mathrm{m}}$ becomes an increasingly large fraction of the total rainfall, and mesoscale downdraughts

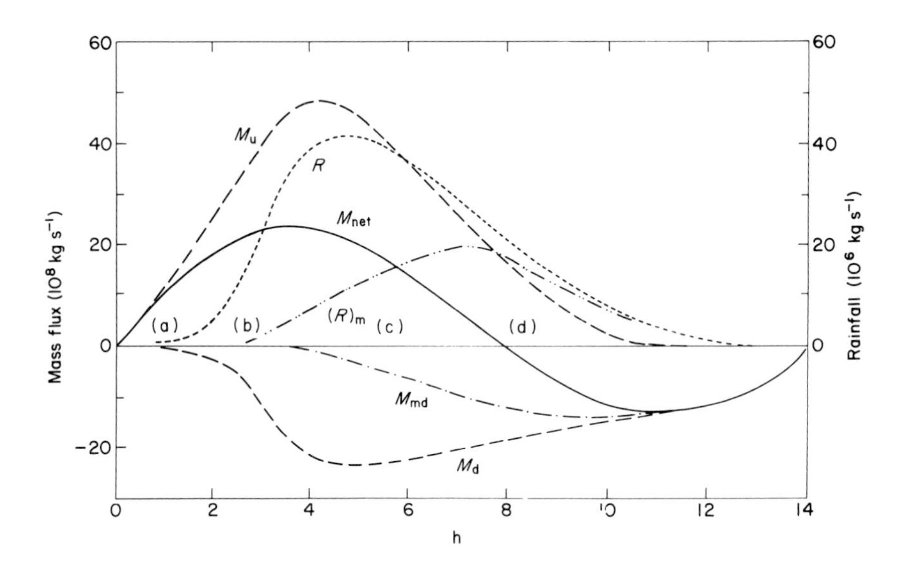

Fig. 8. Schematic time series illustrating life-cycle of "typical" mesoscale convective system reaching a length and width of 200 km (after Zipser, 1980).

Stages of life cycle: (a) = formative; (b) = intensifying; (c) = mature; (d) = dissipating (Leary and Houze, 1979); - - - - -, R = total rainfall rate; —··—, $(R)_m$ = rainfall rate from anvil; ———, M_{net} = net mass flux in lowest km ($= M_u - M_d$), — — —, $M_u (M_d)$ = upward (downward) mass flux through 1-km level – includes *organized* convective draughts; — · —, M_{md} = downward mass flux in mesoscale downdraughts.

(M_{md}) become an increasingly large fraction of the total downdraught mass flux. It is important to recognize that the net mass flux becomes negative during the dissipating stage of the system, signifying not that upward motion has ceased or that convection has ceased, but that the downdraught part of the circulation system now dominates.

Confusion can arise if the life-cycle aspects are ignored, because it is not uncommon for the dissipating stage to occupy half the total lifetime. This stage also contains the greatest areal extent of high clouds, and therefore the most prominent disturbance from a satellite perspective, and if the stratiform precipitation region is electrically active (note Figs. 2, 3) the greatest number of surface observing stations will be reporting "thunderstorms". Yet it is precisely at this time that the correct forecast will often be for a complete cessation of precipitation within a few hours.

4 Impact on nowcasting

Mesoscale convective systems of the type defined and described here are dominant in the tropics, and common in mid-latitudes. Maddox (1980) gives statistics for the occurrence of mesoscale convective complexes, and despite his particular definition, which includes only very large systems of this type, 43 were counted in the midwestern United States during the warm season of 1978. By any definition, such systems are involved directly in some of the most difficult and pernicious forecasting situations known to the profession.

The life-cycle model proposed here is conceptual. Its beneficial use depends upon recognition of an evolving situation and cautious application of the conceptual model to arrive at a conclusion. The tasks are first one of analysis (i.e. does the model apply to this situation, and how are current observations to be interpreted), and only then one of very-short-range forecasting.

A familiar example of a conceptual life-cycle model in meteorology which has been useful is the "Norwegian Wave Cyclone Model". Prior to the days of dynamical weather prediction, forecasters learned to anticipate situations when wave development was likely along fronts, and at least equally important, they learned to recognize observational evidence that potential development was, in fact, taking place. They also learned to recognize when an intensifying wave cyclone was about to go through the occlusion process, which was helpful in anticipating the weather distribution about the cyclone as well as its imminent decay phase.

Analogously, there are no dynamical prediction models which can be used now which are appropriate for very-short-range prediction of mesoscale convective systems. The potential for application of conceptual models clearly exists, but more experience is required before specific forecast rules can be proposed. A first step could be the application of such a life-cycle model to a large number of mesoscale convective systems, to find out whether the insights which seem to have been gained in the analysis of cases on a research basis (previous section) will remain valid. In order to make progress in this direction, it is necessary to observe mesoscale weather systems adequately, which requires appropriate radar data and mesoscale surface data as an absolute minimum, and in most situations data from an upper-air sounding network and satellite data are also required. As experience is gained, it should become possible to devise procedures for analysis which could lead to a forecaster anticipating the formation and intensification of a mesoscale convective system, and the decay of existing ones. An indicator of decay might be the passage of a strong mesoscale system through its mature stage in the absence of strong continued large-scale forcing.

Potential areas of improvement of very-short-range forecasts from eventual application of such conceptual models might include:

1. Narrowing down the area where new convective cloud growth is likely

to the leading edge. In most cases this would circumscribe the area of potential severe weather or very heavy rainfall. Recent history of quantitative radar data and of low-level winds would be necessary in order to define the area of largest potential convergence of warm, moist air into the system.

2. Anticipation of the development of the stratiform rain area could lead to an important distinction in the forecast between continued risk of severe weather, and liklihood of several hours of light-to-moderate steady rain.

3. Anticipation of the cessation of extensive stratiform or anvil-type rain could lead to a forecast of the ending of all precipitation over most of the area currently being affected by the dissipating mesoscale system. In many circumstances this might be the forecasting advance most likely to be made in the near future, if recognition of dissipating systems proves to be routinely possible in actual practice. Often, forecasters are reluctant to depart from a "scattered thunderstorms" forecast, especially when rain and thunder continue to be reported in the area, but when the most likely outcome is "rain ending, followed by gradual clearing", the potential for improvement in the product is considerable.

Acknowledgements

The author is most appreciative of the contributions of material, ideas, and helpful discussions with Alan Betts, John M. Brown, Keith Browning, J. Michael Fritsch, Robert Houze, Margaret LeMone, and Robert Maddox.

References

Austin, G. L. and Bellon, A. (1982). This volume, pp. 177–190.

Betts, A. K. (1978). Convection in the tropics. *In* Meteorology over the Tropical Oceans, Royal Meteorological Society, Bracknell, U.K. pp. 105–132.

Browning, K. A. (1977). The structure and mechanisms of hailstorms. *In* Met. Monogr. No. 38, American Meteorological Society.

Fritsch, J. M., Chappell, C. F. and Hoxit, L. R. (1976). The use of large-scale budgets for convective parameterization. *Mon. Weath. Rev.* **104**, 1408–1418.

Gamache, J. F. and Houze, R. A., Jr. (1982). Mesoscale air motions associated with a tropical squall line. *Mon. Weath. Rev.* **110**, 118–135.

Hamilton, R. A. and Archbold, J. W. (1945). Meteorology of Nigeria and adjacent territory. *Q. J. R. met. Soc.* **71**, 231–262.

Houze, R. A., Jr. (1977). Structure and dynamics of a tropical squall line system observed during GATE. *Mon. Weath. Rev.* **105**, 1568–1589.

Houze, R. A., Jr. (1981). Structures of atmospheric precipitation systems – A global survey. *Radio Sci.* **16**, 671–689.

Houze, R. A., Jr. and Betts, A. K. (1981). Convection in GATE. *Rev. Geophys. Space Phys.* **19**, 541–576.

Leary, C. A. and Houze, R. A. Jr., (1979). The structure and evolution of convection in a tropical cloud cluster. *J. atmos. Sci.* **36**, 437–457.

Maddox, R. A. (1980). Mesoscale convective complexes. *Bull. Am. met. Soc.* **61**, 1374–1387.

Newton, C. W. (1963). Dynamics of severe convective storms. Met. Monogr. **5**, 33–58.

Ogura, Y. and Chen, Y.-L. (1977). A life history of an intense mesoscale convective storm in Oklahoma. *J. atmos. Sci.* **34**, 1458–1476.

Ogura, Y. and Liou, M.-T. (1980). The structure of a mid-latitude squall line: A case study. *J. atmos. Sci.* **37**, 553–567.

Pedgley, D. E. (1962). A meso-synoptic analysis of the thunderstorms of 28 August 1958. Geophysics Memoirs No. 106, Meteorological Office, London, UK

Sanders, F. and Emanuel, K. A. (1977). The momentum budget and temporal evolution of a mesoscale convective system. *J. atmos. Sci.* **34**, 322–330.

Zipser, E. J. (1969). The role of organized unsaturated convective downdrafts in the structure and rapid decay of an equatorial disturbance. *J. appl. Met.* **8**, 799–814.

Zipser, E. J. (1977). Mesoscale and convective-scale downdrafts as distinct components of squall-line circulation. *Mon. Weath. Rev.* **105**, 1568–1589.

Zipser, E. J. (1980). Kinematic and themodynamic features of mesoscale systems in GATE. Proceedings of Seminar on Impact of GATE on Large-Scale Numerical Modeling of the Atmosphere and Ocean. Woods Hole, Massachusetts, August 1979, pp. 91–99. National Academy of Science, Washington, D. C.

Zipser, E. J., Meitin, R. J. and LeMone, M. A. (1981). Mesoscale motion fields associated with a slowly moving GATE convective band. *J. atmos. Sci.* **38**, 1725–1750.

PART 4

Use of Numerical Models

Introduction

We have seen in earlier chapters that great advances are being made in our ability to observe and interpret the behaviour of the atmosphere on the mesoscale. It would be natural to assume that an immediate way forward would be to incorporate the new nowcast data as input to a mesoscale model. Unfortunately, good though these observations are in many respects, the majority of them are not in the form of a 3-D coverage of dynamically orientated variables such as temperature or wind. Moreover, in order to reconcile the nowcast data with dynamically consistent background fields, a surprising amount of the information content of the nowcast would be lost. Therefore, the most rapid progress in the application of mesoscale models to very-short-range forecasting is to be expected in those circumstances where the initial state of the atmosphere does not need to be specified in detail. These circumstances arise when the predominant forcing is by terrain features such as land–water boundaries and hills. This is the theme developed by Pielke (Chapter 4.1). The theme is extended by Carpenter (Chapter 4.2) to include circumstances where the model can be expected to respond realistically to local forcing due to well-defined mesoscale singularities such as differential surface heating across a major cloud boundary. In these cases the problem is to identify the salient features in the nowcast and to use them to constrain the model rather than to try to use detailed fields of mesoscale observations as input.

The prospect is less encouraging for the early application of mesoscale models to forecasting synoptically induced mesoscale systems, for example squall lines and other mesoscale precipitation systems. In such cases there may be no escaping the need to use detailed mesoscale observations to initialize the models. In the final chapter (4.3) Warner and his colleagues present a case study that illustrates the sensitivity of precipitation forecasts to the use of input from satellite imagery and current surface precipitation. By carefully ensuring the dynamical consistency of the input data, they achieve significant improvements in the model forecasts, but only during the first 4 hours of the forecast. Even then it is still difficult for such models to out-perform forecasts based on simple persistence during the first few hours.

Clearly much research remains to be done in this area. We await with particular interest the outcome of studies of the impact on regional and mesoscale models of high-resolution satellite sounding data of the kind discussed in Chapters 2.3 and 2.4.

K. A. Browning

4.1

The Role of Mesoscale Numerical Models in Very-short-range Forecasting

ROGER A. PIELKE

1 Introduction

During the last 30 years or so, there have been major advances in our ability to forecast weather on time scales from 12 to 48 hours or more into the future over 40 000 km^2 areas in the United States (American Meteorological Society, 1979). This has been achieved by the use of synoptic-scale numerical models and model output statistics derived from such models (e.g. see Pielke, 1977; Klein, 1978). On the synoptic scale, the wind field above the planetary boundary layer is seldom far from gradient wind balance which has aided in providing effective forecasts on that scale. Unfortunately, synoptic-scale forecasts are unable to provide accurate predictions of smaller-scale atmospheric features. These small features include such extreme weather events as hurricanes and thunderstorms producing flash floods and tornadoes.

This smaller scale is generally referred to as the *mesoscale* (e.g. see Orlanski, 1975). Pielke (1981) has recently proposed a dynamical definition of the mesoscale as those atmospheric circulations which have horizontal scales large enough to be essentially in hydrostatic balance, yet small enough so that the wind field deviates substantially from gradient wind balance above the planetary boundary layer. Even smaller atmospheric features, which have a significant non-hydrostatic component, include convective-scale and microscale systems. Martin (1981) examined the horizontal scales for which the hydrostatic assumption is valid in linear and non-linear sea-breeze models and found that greater vertical thermodynamic stability, and/or decreased vertical turbulent mixing of heat, reduced the magnitude of the non-hydrostatic effect.

The mesoscale can also be defined more qualitatively as having a temporal and horizontal spatial scale smaller than the conventional rawinsonde network, but significantly larger than the size of individual cumulus clouds. This implies that the horizontal scale is of the order of a few kilometres to a few hundred kilometres, with a time scale of about one hour to one day. The vertical scale extends from tens of metres to the depth of the troposphere. Scales smaller than this – the convective scales – correspond to atmospheric features which, for weather forecasting purposes, can only be described statistically, while larger systems – the synoptic scale – correspond to those which can be resolved from conventional synoptic observations.

Mesoscale systems can be categorized into those which are primarily forced by the ground surface (i.e. *terrain-induced mesoscale systems*) and those which are substantially driven by propagating larger-scale features (i.e. *synoptically forced mesoscale systems*). Terrain-induced systems include land- and sea-breezes, urban circulations, lake effect storms, mountain-valley circulations, and forced airflow over mountains, while synoptically forced features consist, for example, of squall lines, convective bands embedded in stratiform cloud systems, and hurricanes.

Unfortunately, these latter systems are particularly hard to forecast since the synoptic circulations which generate them are themselves moving and changing in intensity with time. Although terrain can influence their propagation and development, their geographic location is not fixed, since their main reason for existing is because of synoptic or smaller-scale travelling disturbances in the atmosphere.

Terrain-induced mesoscale systems, on the other hand, are primarily forced by fixed topography. Although synoptic conditions can determine the specific spatial and temporal pattern of this type of mesoscale system, its primary reason for existence is a result of terrain inhomogeneities. Therefore, terrain-induced mesoscale systems are much more amenable to accurate predictions, since terrain characteristics are much more easy to determine than are the spatial characteristics of a travelling large-scale disturbance. The application of our advanced understanding of these particular systems, which recur frequently in the same location, offers the best hope for improved very-short-range forecasts for several hours ahead.

In terms of lives lost to particular mesoscale circulations, it is also terrain-induced systems which are most important. As reported by Pielke (1978), on the average, about 600 lives in the United States are lost to squall line and hurricane-related weather hazards (e.g. tornado, storm surge, etc.), while roughly 15 000 lives (as estimated by the United States Academy of Science, HEW, 1976) are lost due to outside air pollution, whose concentration is partially dependent on the convergent/divergent patterns caused by such terrain-induced features as urban–rural circulations, mountain–valley flows and sea- and land-breezes.

2 Terrain-induced mesoscale systems

Sea- and land-breezes, mountain–valley winds, and urban circulations are primarily forced by a horizontal temperature gradient, created as a result of horizontal variations in ground surface characteristics. In the case of sea- and land-breezes, this variation is primarily caused by the different surface heating rates of ground and water, while, with urban circulations, this spatial gradient is caused by large variations in aerodynamic roughness and thermal properties between urbanized and rural areas. Mountain–valley winds result from different heating and cooling rates between the mountain and the free atmosphere at the same altitudes. Figures 1a–1c illustrate examples of wind fields predicted by mesoscale models for each of these systems during summer afternoons in which the prevailing synoptic flow is light. The perturbation of the synoptic flow by the sea breeze (Fig. 1a) and the drainage winds (Fig. 1c) in these experiments is particularly pronounced. In Fig. 1a, a 100-km line is drawn in order to represent the approximate horizontal grid interval used at about 40°N in the United States National Weather Service Limited Fine Mesh (LFM) model. Since at least four grid intervals are needed to resolve adequately a feature in a numerical simulation, it is quite evident that existing synoptic models are unable to resolve even the largest of this type of terrain-induced mesoscale system.

Figures 2a and 2b present examples of meteorological fields associated with a lake effect storm and forced airflow over a terrain feature. As in Fig. 1a, the 100-km line drawn on Fig. 2a corresponds to the approximate grid mesh used in the United States National Weather Service LFM model. The terrain-induced systems exemplified in Figs. 2a and 2b differ from sea- and land-breezes, mountain–valley winds and urban circulations in that they are not primarily forced by differential diurnal heating rates across a region. With lake effect storms, it is the intense turbulent sensible and latent heat flux from large water bodies into the planetary boundary layer as arctic air transits a region that generates localized heavy snow squalls. Such snowfalls are maximized by roughness and elevation changes which cause regions of enhanced low-level convergence on the windward shore. Lavoie (1972) simulated such a lake-effect storm over Lake Erie during a 30-hour period in the winter of 1966, and found excellent agreement between the observed and predicted amounts and pattern of snowfall, as illustrated in Fig. 2a.

Forced airflow over rough terrain also causes large local variations in meteorological properties as air is forcibly lifted over a topographic barrier. If the air is sufficiently moist and the ascent is to great enough heights, condensation and/or sublimation occurs on the windward slopes, while dry, adiabatic descent usually occurs over most of the lee slopes. Precipitation patterns are often substantially influenced by this orientation of the terrain with respect to the wind (e.g. Rhea, 1977; Collier, 1977). In addition, damaging chinook winds occasionally result just downwind of such topographic barriers (e.g. see Lilly and Zipser, 1972). The three-dimensional simulation by Clark (1981), for example, provided a reasonable simulation of the asymmetric wind flow in the vicinity of Elk Mountain, Wyoming, with stronger winds to the north of the highest terrain, as shown in Fig. 2b.

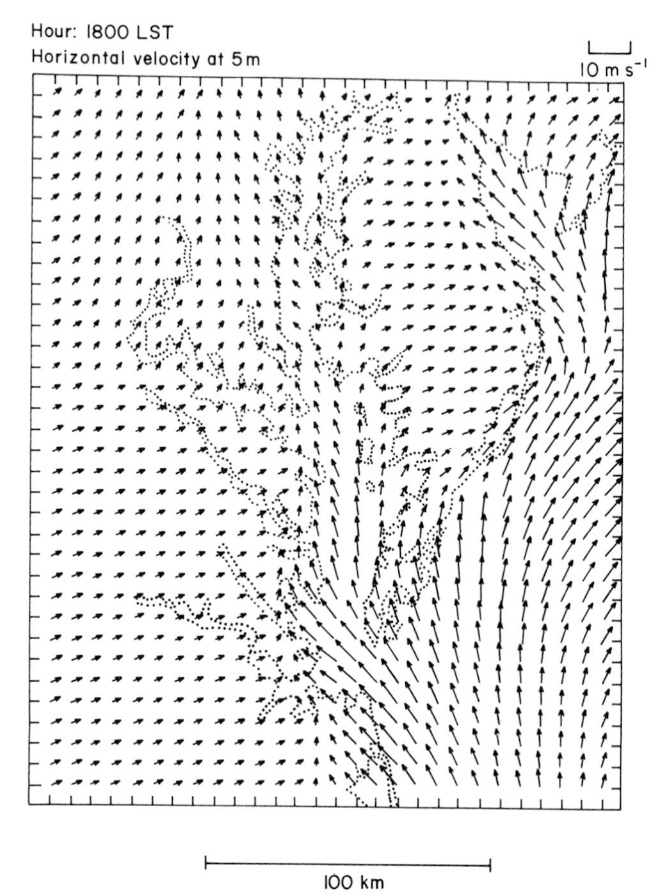

Fig. 1a. The predicted winds at 5 m at 1800 LST on 21 July 1978 over the Chesapeake Bay region (from Segal and Pielke, 1981). The verification of the predicted meteorological fields for this date is reported in Segal and Pielke, and Segal *et al.* (1981). Among the results, the veering of the surface wind along the coast due to diurnal heating of the ground was realistically predicted by the model. This heating was accurately simulated. The root mean square error for the 2 m temperature for 12 observation sites, for example, was 1.1°C, while the standard deviations of the observations and the predictions were 2.1 and 2.2°C – a result indicative of positive model skill.

Pielke (1981) presents a more extensive review of terrain-induced mesoscale systems . But, even for the few examples illustrated here in Figs. 1a–1c, 2a and 2b, it is obvious that existing weather service models (even those defined as "fine mesh") would be unable to resolve these smaller-scale atmospheric features. For the examples presented in these figures, horizontal grid intervals from 1 km (Fig. 1c) to about 10 km (Fig. 1a) are required.

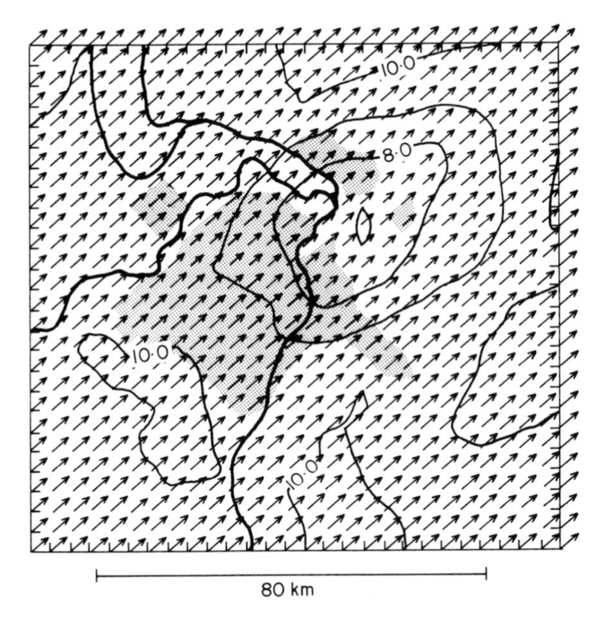

Fig. 1b. The predicted winds at 1 km at 1100 LST on 18 July 1975 over St Louis, Missouri. The urban land use areas are located within the stippled section. The wind arrows are drawn to a scale of 8 m s^{-1} per grid interval, while the isotachs are contoured at 1 m s^{-1} intervals. Although some direction change is noted downwind of the urban area, most of the velocity perturbation is in wind speed (from greater than 10 m s^{-1} upwind to 7 m s^{-1} just downwind from the city). Observations from eight pibal sites for this time showed wind speeds at 1.25 km of almost 10 m s^{-1} upwind and less than 9 m s^{-1} downwind of the city in essentially the same locations as predicted by the model (from Hjelmfelt, 1980; observations extracted by Hjelmfelt from Achtemeier, 1980).

The size of forecast zones actually used in the routine dissemination of weather predictions illustrates the emphasis of weather services on synoptic-scale weather systems as opposed to mesoscale features. Figure 3, for example, illustrates the size of the ten United States Weather Service forecast zones in Virginia, for which at least four times a day, area forecasts are distributed to the public. Little, if any, mention is ever made of variations of weather within these individual zones.

The results from mesoscale observations and numerical model simulations, therefore, yield the conclusions that,

- significant terrain-induced atmospheric circulations occur on scales too small to be resolved by operational weather service numerical prediction models; and

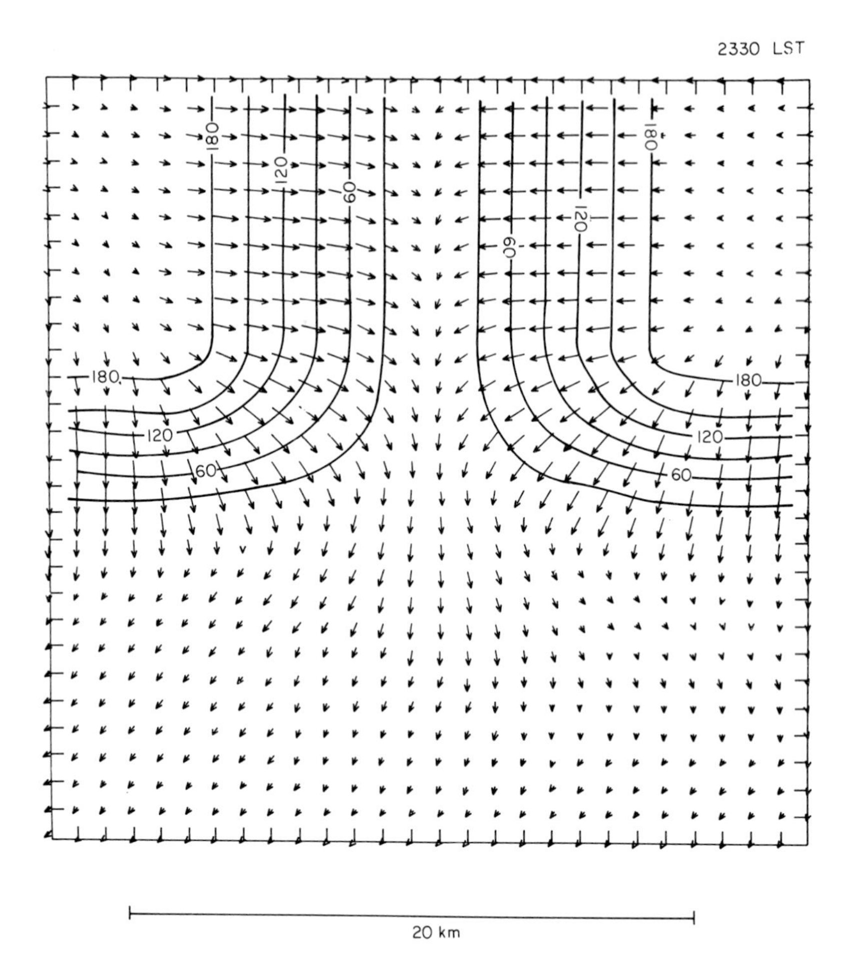

Fig. 1c. The predicted winds at 2330 LST at 8 m above the terrain for an idealized valley 19 km wide and 2 km deep. A wind vector of one grid interval in length corresponds to $8\ \mathrm{m\,s^{-1}}$. Terrain contours are plotted at 30-m intervals. Although not verified from specific observations, the predictions in this and other spatial cross-sections (not shown) illustrate shallow-slope flow off of the valley walls with a deeper out-valley flow emanating from over the centre of the valley (from McNider, 1981).

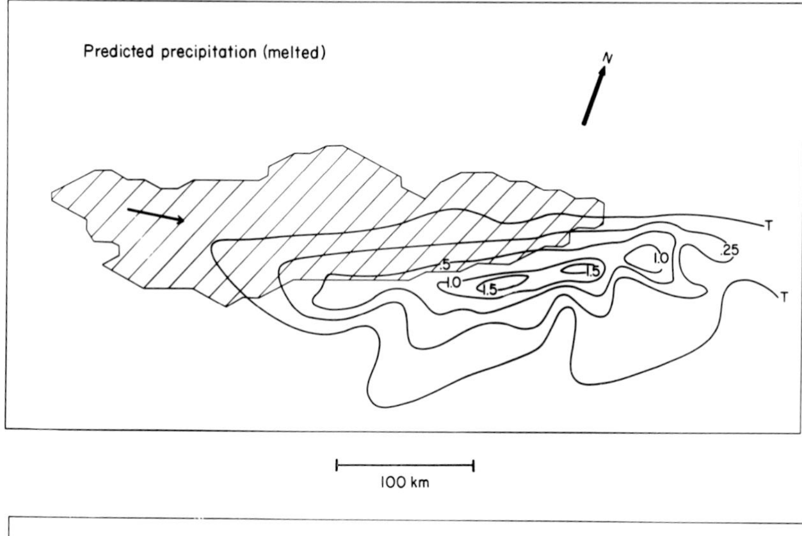

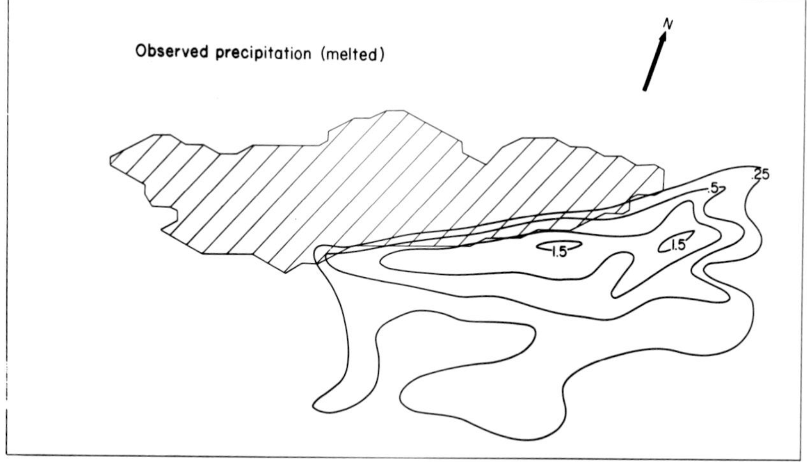

Fig. 2a. The predicted (top) and observed (bottom) precipitation (melted) in cm for the 30-h lake effect snowstorm of 1–2 December 1966 (from Lavoie, 1972).

- mesoscale numerical models can realistically simulate these meteorological systems.

3 Two ways of using models for forecasting terrain-induced mesoscale systems

Satellite and radar, with supplemental surface observations, provide the most effective tools for accurate nowcasts. Unfortunately, except by using simple

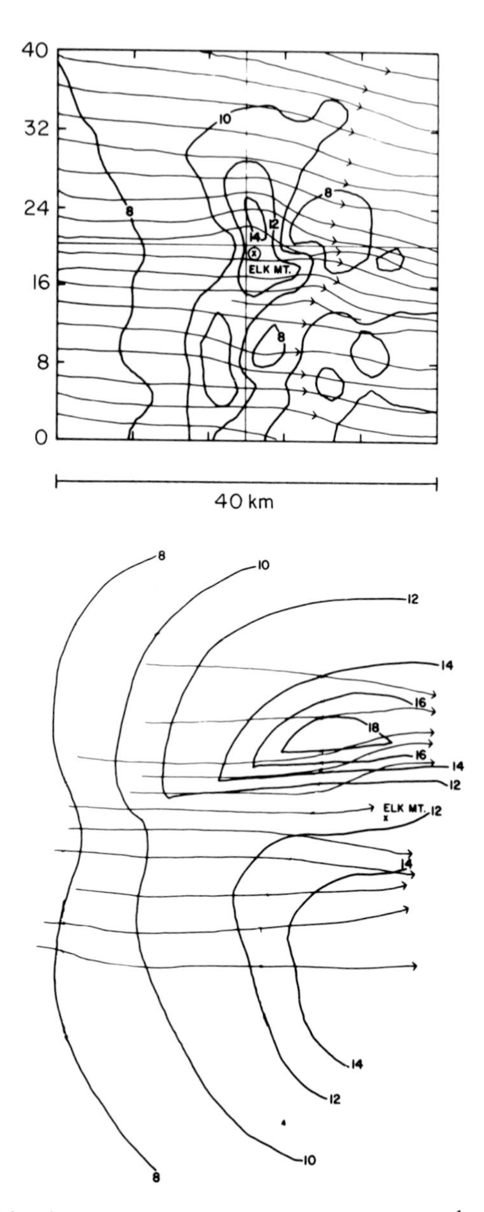

Fig. 2b. Predicted (top) and observed (bottom) isotachs in m s^{-1} and streamlines on 22 December 1976, at 1.36 km above the flat ground due to the presence of a ~ 1.3-km high mountain (Elk Mountain, Wyoming) and nearby elevated terrain (from Clark and Gall, 1981, where they used observed data reported in Karacostas and Marwitz, 1980). Both figures have approximately the same spatial scale.

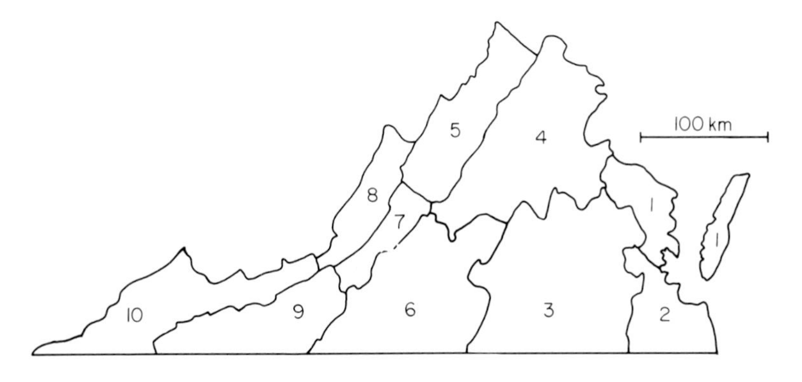

Fig. 3. National weather forecast zones in Virginia in 1980.

Virginia zone forecast areas: 1, East-central coastal; 2, Southeastern coastal; 3, Southeastern Piedmont; 4, Northeastern Piedmont; 5, Northern Shenandoah Valley; 6, Southwestern Piedmont; 7, Southern Shenandoah Valley; 8, Central mountains; 9, Southwestern plateau; 10, Southwestern mountain.

forward extrapolation, forecasts (out to 12 to 24 hours) are not possible using an observation platform alone. On the other hand, very-short-range predictions (less than 2 hours) with a mesoscale model are not practical since it takes time to assimilate the initialization data, run the model, and to disseminate the results. Therefore, models are most effective for time periods beyond 2 hours.

Mesoscale models can be utilized in order to upgrade very-short-range weather forecasts in either one of the two following modes:

- a mesoscale model can be integrated one or more times daily in the same fashion as the LFM model; or
- a series of mesoscale model runs can be archived for typical synoptic situations and then referred to for forecast guidance as needed.

The first methodology is the more desirable since initial and boundary conditions corresponding to specific dates and times can be included. Larger-scale models (e.g. the LFM) could be used for initialization, with changes in the synoptic fields included as performed by Carpenter (1979). As with synoptic models, the method of Model Output Statistics (MOS), such as discussed by Klein (1978), could be used to statistically improve the raw mesoscale numerical model output.

With synoptically induced mesoscale systems, this is the only procedure that is reasonable, since the synoptic forcing is highly variable in timing and structure, and seldom, if ever, recurs in the same form. The resultant mesoscale systems, therefore, seldom repeat with the same spatial and temporal structure.

With terrain-induced mesoscale systems, however, the forcing by the ground is often very similar from day to day, so that similar local weather may recur frequently. The difference between days is expected to be due to variations in

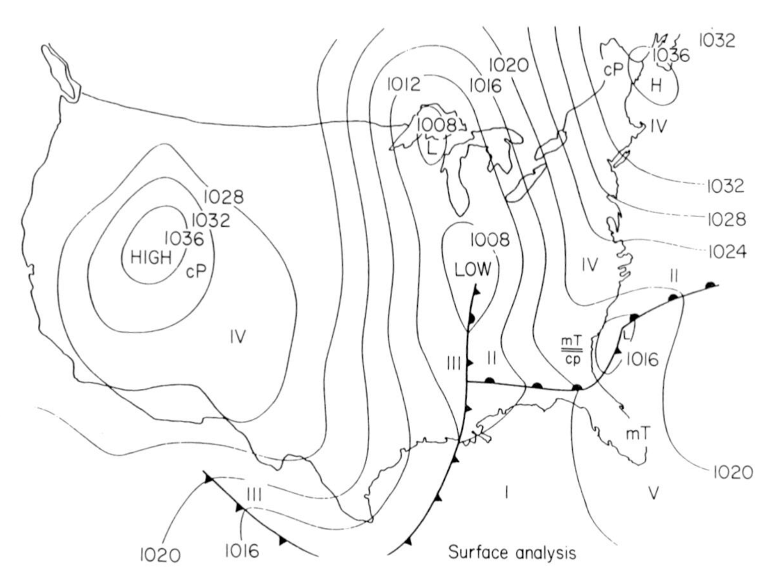

Fig. 4. The classification of a surface analysis chart for five synoptic categories for 9 January 1964 (from Lindsey, 1980, and Garstang *et al.,* 1979).

the synoptic flow over the terrain, which influences the specific spatial and temporal patterning of the mesoscale meteorological variables.

A procedure for classifying typical synoptic situations has recently been reported in Lindsey (1980) and Garstang *et al.* (1979). In this scheme, as illustrated using the five numbered categories drafted on Fig. 4, the position of a site relative to the polar front and associated extratropical storms is determined. The primary reason for this classification is that each location is expected to have a different characteristic thermal stratification and wind profile. These categories, whose derivation makes use of the omega equation (e.g. see Holton, 1972), are summarized in Table 1.

As reported in Lindsey's thesis, seasons can be defined in terms of the percentage of time a given station is north of the polar front (categories II, III and IV), rather than south of the polar front (categories I and V). In addition, Lindsey determined the wind rose for each category during each season along with the average speed for each direction.

Using information of this type, it is straightforward to perform a number of mesoscale model calculations in specific geographic regions for the most frequently occurring synoptic categories during a particular season. Average values of the lower tropospheric wind speed and direction, and of the temperature lapse rate, can be used as the synoptic input, while ground surface characteristics representative of that season are used to determine the bottom boundary

TABLE 1

Synoptic classification scheme used when integrating a mesoscale model using climatological data (modified from Lindsey, 1980)

Category	Air mass	Reason for categorization[a]
I	*mT*	*In the warm sector of an extratropical cyclone.* In this region the thickness and vorticity advection is weak with little curvature to the surface isobars. There is limited low level convergence with an upper level ridge tending to produce subsidence. Southerly low-level winds are typical
II	mT/cP, mT/cA, mP/cA	*Ahead of the warm front in the region of cyclonic curvature to the surface isobars.* Warm air advecting upslope over the cold air stabilizes the thermal stratification, while positive vorticity advection and low-level frictional convergence add to the vertical lifting. Because of the warm advection, the geostrophic winds veer with height. Low-level winds are generally north-easterly through south-easterly
III	*cP*; *cA*	*Behind the cold front in the region of cyclonic curvature to the surface isobars.* Positive vorticity advection and negative thermal advection dominate, with the resultant cooling causing strong boundary layer mixing. The resulting thermal stratification in the lower troposphere is neutral, or even slightly, superadiabatic. Gusty winds are usually associated with this sector of an extratropical cyclone. Because of the cold advection, the geostrophic winds back with height. Low-level winds are generally from the north-west through south-west
IV	*cP*; *cA*	*Under a polar high in a region of anticyclonic curvature to the surface isobars.* Negative vorticity, weak negative thermal advection and low-level frictional divergence usually occur, producing boundary layer subsidence. Because of relatively cool air aloft, the thermal stratification is only slightly stabilized during the day, despite the subsidence. At night, however, the relatively weak surface pressure gradient associated with this category causes very stable layers near the ground on clear nights due to long-wave radiational cooling. The low-level geostrophic winds are usually light to moderate varying slowly from north-westerly to south-easterly as the ridge progresses eastward past a fixed location
V	*mT*	*In the vicinity of a subtropical ridge* where the vorticity and thickness advection, and the horizontal pressure gradient at all levels are weak. The large upper-level ridge, along with the anticyclonically curved low level pressure field, produces weak but persistent subsidence. This sinking causes a stabilization of the atmosphere throughout the troposphere. Low-level winds over the eastern United States associated with these systems tend to blow from the south-east through south-west

[a]This discussion applies to northern hemisphere.

conditions. Values of wind speed and direction, and lapse rate within one standard deviation of the observed synoptic mean value could be used to provide a representative sample of expected terrain-induced mesoscale patterns.

Garstang *et al.* (1980) and Snow (1981) reported on the use of a mesoscale model in this fashion in order to estimate wind power potential over 10^5 km^2 areas along three coastal sections of the Gulf and Atlantic coasts of the United States. The primary forcing for each region was the temperature contrast between land and water. Most of the remaining difference in the mesoscale pattern between days is due to different categories of larger-scale flow over each region.

Using three different model runs at each location, in order to represent the three most typical synoptic categories within each coastal section, and weighing the results according to the frequency of occurrence of each particular synoptic category, good agreement was obtained between model estimated and long-term observed climatological wind data for two of the three regions studied. One region (the Texas coast) was not well simulated, but this was only because the terrain-induced mesoscale circulation caused by the sloping terrain in that area was neglected in the model runs, as discussed by Snow (1981). For the three sites, for 60%, 75% and 78% of the time, the large-scale conditions were determined to be reasonably well represented by using just three different synoptic categories (Garstang *et al.*, 1980).

It appears, therefore, that for improved local forecasts, only a limited number of mesoscale predictions need be made if the atmospheric system is primarily terrain-induced, and if the frequency of synoptic patterns can be determined.

4 Conclusions

Models exist today which can provide effective very-short-range forecasts of terrain-induced mesoscale systems. For periods less than two to four hours or so, the use of radar and satellite imagery, along with simple very-short-term extrapolation, probably offer the best procedure. For longer time periods, mesoscale model predictions, either calculated in real time or archived for typical synoptic conditions, should be used. Such models currently have a physical and numerical sophistication equal to that of the synoptic models in operational use.

Further research is needed, however, before synoptically induced mesoscale systems, such as squall lines and hurricanes, can be accurately modelled in real time, since the initial and boundary conditions required for such simulations are highly variable in time and space, and seldom, if ever, recur in the same pattern. Since travelling larger-scale systems are the primary forcing of this type of mesoscale feature, they must be very accurately represented on the mesoscale model grid. The cost of such simulations would be of the same order as required for synoptic-scale models. Fortunately, costs of computer systems should soon be low enough so that each state in the United States could afford to run such models in real time, although the serious problem of obtaining accurate observed input data remains.

Terrain-induced mesoscale systems, on the other hand, can be simulated using *climatologically* expected synoptic conditions for the initialization and spatial boundary conditions. In contrast to the synoptically induced circulation, these systems are primarily forced by terrain inhomogenieties, which are more or less the same over long periods of time. After the initial integration of a mesoscale model for the range of most likely synoptic categories, the cost of this approach is very low. Dissemination problems would be minimized since a booklet or computer tape of expected mesoscale conditions would be distributed to a wide variety of users. When synoptically induced and terrain-induced systems interact, it may be possible to superimpose radar and satellite observations onto the mesoscale climatological forecasts, as proposed by Pielke (1977), in order to obtain an inexpensive estimate of short-term subsequent weather.

Acknowledgements

The author wishes to acknowledge the Atmospheric Science section of the National Science Foundation, the Computing Facility of the National Center for Atmospheric Sciences, and the American Geophysical Union, for providing support to perform portions of the work reported in this paper. The typing and useful editing advice was competently handled by Ann Gaynor and Lois Bond and their support is gratefully appreciated.

References

Achtemeier, G. L. (1980). Boundary layer structure and its relation to precipitation over the St. Louis area. Illinois State Water Survey Report No. 2 under NSF Grant ATM8-08865.

American Meteorological Society (1979). "Policy Statement of the American Meteorological Society on Weather Forecasting" as adopted by the Council on September 27, 1979. *Bull. Am. met. Soc.* **60**, 1453–1454

Carpenter, K. M. (1979). An experimental forecast using a non-hydrostatic mesoscale model. *Q. J. R. met. Soc.* **105**, 629–655.

Clark, T. L. and Gall, R. (1981). Three-dimensional numerical model simulations of airflow over mountainous terrain: A comparison with observations. *Mon. Weath. Rev.* In press.

Collier, C. G. (1977). The effect of model grid length and orographic rainfall efficiency on computed surface rainfall. *Q. J. R. met. Soc.* **103**, 247–253.

Garstang, M., Pielke, R. A. Gusdorf, J., Lindsey, C. and Snow, J. W. (1979).

Coastal zone wind energy, Part I: Synoptic and mesoscale controls and distributions of coastal wind energy. DOE/ET/2027Y-1, NTIS, Springfield, Virginia, 187 pp.

Garstang, M., Pielke, R. A. and Snow, J. W. (1980). Coastal zone wind energy, Part I: Potential wind power density fields based on 3-D simulations of the dominant wind regimes for three East and Gulf coast areas. Draft Final Report. Department of Energy, Washington, D.C. 20545, 31 pp.

HEW (1976). Human health and the environment. Some research needs. NIH-77-1277, Department of Health, Education and Welfare, U.S. Government Printing Office, Washington, D.C. 20401.

Hjelmfelt, M. (1980). Numerical simulation of the effects of St. Louis on boundary layer airflow and convection. Ph.D. dissertation, University of Chicago, Chicago, Illinois, 60680, 158 pp.

Holton, J. R. (1972). "An Introduction to Dynamic Meteorology", 319 pp. Academic Press, New York and London.

Karacostas, T. S. and Marwitz, J. D. (1980). Turbulent kinetic energy budgets over mountainous terrain. *J. appl. Met.* **19**, 163–174.

Klein, W. H. (1978). Objective forecasts of local weather by means of model output statistics. *In* "Seminars 1978: The Interpretation and Use of Large-scale Numerical Forecast Products", pp. 186–220. European Centre for Medium Range Weather Forecasts. Shinfield Park, England.

Lavoie, R. L. (1972). A mesoscale numerical model of lake-effect storms. *J. atmos. Sci.* **29**, 1025–1040.

Lilly, D. K. and Zipser, E. J. (1972). The front range windstorm of 11 January 1972 – A meteorological narrative. *Weatherwise* **25**, 56–63.

Lindsey, C. G. (1980). Analysis of coastal wind energy regimes. M.S. Thesis, Department of Environmental Sciences, University of Virginia, Charlottesville, Virginia 22903, 183 pp.

Martin, C. L. (1981). Numerical accuracy in a mesoscale meteorological model. M.S. Thesis, Department of Environmental Sciences, University of Virginia, Charlottesville, Virginia 22903, 85 pp.

McNider, R. T. (1981). Investigation of the impact of topographic circulations on the transport and dispersion of air pollutants. Ph.D Thesis, Department of Environmental Sciences, University of Virginia, Charlottesville, Virginia 22903.

Orlanski, I. (1975). A rational subdivision of scales for atmospheric processes. *Bull. Am. met. Soc.*, **56**, 527–530.

Pielke, R. A. (1977). An overview of recent work in weather forecasting and suggestions for future work – STAC Scientific Review. *Bull. Am. met. Soc.* **58**, 506–519.

Pielke, R. A. (1978). Air pollution – a national concern, *Bull. Am. met. Soc.* **59**, 1461.

Pielke, R. A. (1978). The role of man and machine in the weather service of the future. Conference Preprint Volume of the American Meteorological Society Conference on Weather Forecasting and Analysis and Aviation Meteorology, October 16–19, Silver Spring, Maryland, pp. 271–272.

Pielke, R. A. (1981). Mesoscale numerical modelling. *In* "Advances in Geophysics", vol. 23, pp. 185–344. Academic Press, New York and London.

Rhea, O. J. (1977). Winter quantitative mountain precipitation forecasting using an orographic precipitation model as an objective aid. Presented at 1977

Annual Meeting of the American Meteorological Society. Portion of Ph.D. Dissertation presented at Colorado State University, Fort Collins, Colorado 80521.

Segal, M. and Pielke, R. A. (1981). Numerical model simulation of human biometeorological heat load conditions – summer day case study for the Chesapeake Bay area. *J. appl. Met.* **20**, 735–749.

Segal, M., McNider, R. T., Pielke, R. A. and McDougal, D. S. (1981). A numerical model study of the regional air pollution meteorology of the Greater Chesapeake Bay area – summer day case study. *Atmos. Environ.* in press.

Snow, J. W. (1981). Wind power assessment along the Atlantic and Gulf Coast of the United States. Ph.D. Dissertation, Department of Environmental Sciences, University of Virginia, Charlottesville, Virginia 22903, 244 pp.

4.2

Model Forecasts for Locally Forced Mesoscale Systems

K. M. CARPENTER

1 Introduction

A local weather forecast model has been under development in the UK Meteorological Office for some years (Tapp and White, 1976) and two full case studies have been completed and reported (Carpenter, 1979; Bailey *et al.*, 1981). The esssential details of these two examples are presented in sections 3 and 4 and discussed in the context of very-short-range forecasting in section 5. The model itself is summarised in section 2. It will be argued that these examples illustrate classes of situations (which might be larger than is obvious at first sight) in which local weather is determined as a locally forced response to large-scale meteorology. Thus these examples support Pielke's (1982) suggestion that models can be expected to give good forecasts in these situations even in the absence of a mesoscale data base to give good initial conditions.

2 The model

The model is based on a finite-difference approximation of the non-hydrostatic compressible equations of motion on a three-dimensional grid. The horizontal grid has a grid length of 10 km and covers England and Wales. There are ten levels and the top level is at 4 km. This grid is not a permanent feature of the model and it might have been natural to increase the model depth for the second study, in which the mesoscale environment of a deep, stationary storm was forecast. However, that could only be done at the expense of resolution and, in the event, the results suggest that there would be little benefit from using a deeper model. The equations are written in terms of the vertical coordinate $\eta = z - E$, where E is terrain height, so the effects of orography are included in the model.

Boundary-layer turbulence is included in the model, using a K-theory approach, and there is a convective adjustment. However, the effects of moisture, cloud and radiation are described only in so far as they affect the exchanges of heat at the surface.

The calculation of surface fluxes of heat and momentum has been described by Carpenter (1977). Monin–Obukhov similarity theory is used, and the surface temperature is given by requiring heat balance at the surface:

$$S + R_{\downarrow} = H + LE + G + R_{\uparrow}, \qquad (1)$$

where S is the solar heating of the surface,
$R_{\downarrow}$ and $R_{\uparrow}$ are the thermal radiation fluxes at the surface,
H is the sensible heat flux into the atmosphere,
E is the flux of water vapour into the atmosphere,
L is the latent heat of vaporization,
and G is the heat flux into the ground.

The calculation of the evaporation E needs a value for humidity mixing ratio, which is otherwise absent from the model, at the bottom level; a constant value of 0.01 has been used. The solar radiation S is specified as a function of time of day using formulae that allow for season, atmospheric attenuation, reflection from the surface and (in the second example) the presence or absence of cloud.

A limited-area atmospheric model needs initial conditions, (i.e. initial values for its variables), which serve to describe the synoptic situation in addition to any mesoscale details that one might hope to infer from observations. In the present studies, no attempt has been made to use observations to improve the initial conditions, which were interpolated from a synoptic-scale model. For many situations, e.g. a very-short-range forecast of the behaviour of a mesoscale convective complex, this lack of detail in the initial conditions would produce a bad forecast, but the use of smooth fields based on large scale forecasts works well in the examples presented here. A limited area model also needs boundary conditions and, again, these were calculated from the synoptic-scale forecast for the period in question, as they would always have to be. Changes in the prevailing large-scale situation are thus expressed in the boundary conditions and the model forecasts can respond to them.

3 Forecast for a sea breeze situation

The model summarised in section 2 has been used to forecast the development and movement of several sea breeze fronts over England on a hot summer day (Carpenter, 1979). Simpson *et al.* (1977) used observations made on 14 June 1973 to illustrate their study of the inland penetration of sea breezes and their analysis provided a good framework within which to test a model forecast. Simpson's surface analysis for 1800 GMT is shown in Fig. 1.

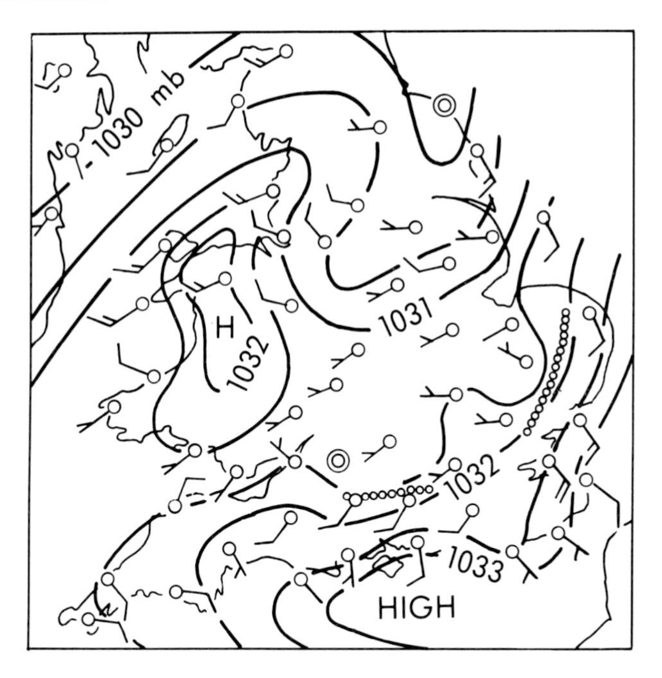

Fig. 1. Observed surface pressure and winds in England and Wales, 14 June 1973 at 18 GMT. The rows of circles show the positions of sea breeze fronts. Note that, although not marked as a front, the observations also suggest a sharp wind shear along the north-east coast. (From Simpson *et al.*, 1977.)

The mesoscale model was run for 24 hours from 0400 GMT, and the sea-level pressures and low level winds forecast for 1800 GMT are shown in Fig. 2. The overall agreement between forecast and analysis is encouraging. The sea breeze fronts are correctly positioned. Moreover the high pressure over central Wales, the troughing along the north-east coast (which is enhanced by the chain of mountains, the Pennines), and the shear line along the north-east coast are all reproduced. The most telling criticism is that the forecast winds are, generally, slightly backed and slightly too strong. This fault becomes more marked later in the forecast, particularly during the night, and is most naturally explained by the fact that the observed winds are much closer to the surface than the forecast (50 m) winds.

Carpenter (1979) discussed this forecast in detail and came to two important conclusions. The first was that the model forecast the position of the sea breeze fronts well at all times except in the morning (when the forecast south coast sea breeze was slow to move inland, possibly because of the coarse vertical resolution). In particular, the observed inland acceleration of the south coast

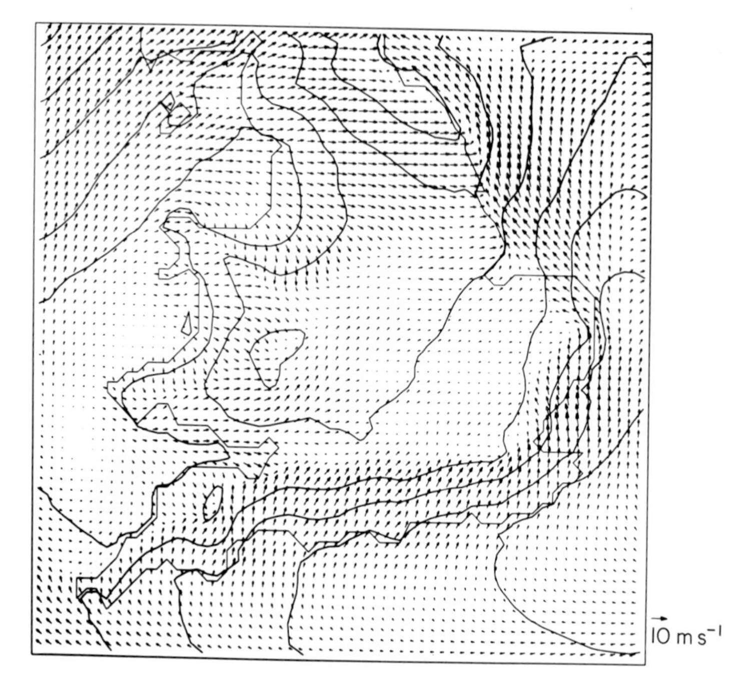

Fig. 2. A 50-m wind and sea-level pressure forecast for 18 GMT on 14 June 1973. The isopleth interval is 0.5 mb.

sea breeze in the evening was reproduced very clearly and accurately. The second was that the inclusion of changes in the boundary conditions due to the movement of the synoptic-scale anticyclone across the country (the large-scale forcing) was critical to this success.

Earlier work by Pielke (1974) had already shown that it was possible to use a general purpose atmospheric model with a grid resolution of about 10 km to simulate the formation and behaviour of sea breezes (in his case, over the Florida peninsular in a variety of situations). The forecast reviewed above showed that models can be used to give accurate sea breeze forecasts for individual days, even when the large-scale forcing is not constant. In Pielke's simulations, the large-scale flow was always simple and constant, but our forecast suggests that these limitations were not necessary to his success. The most interesting general point arising from the success of both models is that, although neither can resolve the detailed structure of the sea breeze front (where sharp variations exist over hundreds or even tens of metres) they still describe the discontinuity in temperature and wind realistically, and forecast the position and movement of the fronts accurately.

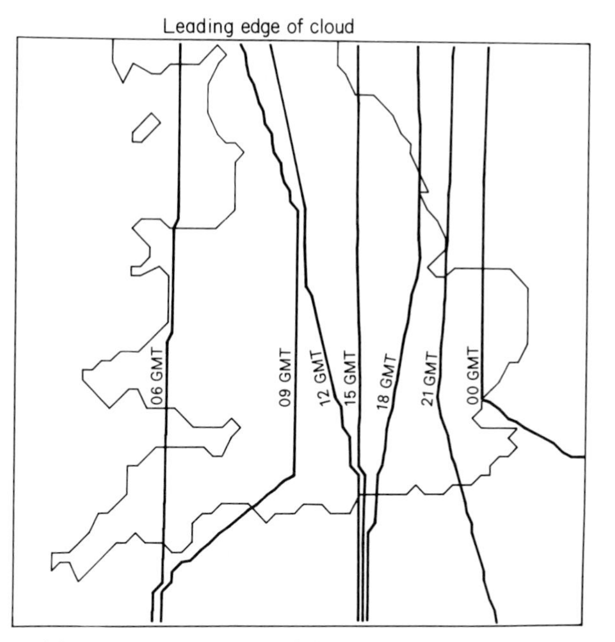

Fig. 3. The position at various times of the leading edge of a layer of cloud that moved across England on 14 August 1975.

4 Forecast for a convective storm situation

On 14 August 1975, a stationary storm centred over northwest London caused extensive flooding. The natural history of this storm, the Hampstead storm, has been well rehearsed (Keers and Westcott, 1976; Atkinson 1977) and the dynamics of the storm itself have been studied using a numerical model (Miller, 1978). With this background, this situation is a natural choice as a case-study for testing a mesoscale forecast model like that summarized in section 2. Such a model should provide detail that is not achievable in a synoptic-scale model and thus help to elucidate or forecast the environment in which strictly local events like the Hampstead storm might occur.

The standard version of the model described in section 2 does not include any treatment of clouds. Fortunately, the most obvious feature of the cloud cover was a bank of altocumulus spreading from the west. With the exception of small areas in the west of the country, once this cloud arrived at any place it remained cloudy there for the rest of the day. Thus it was possible to allow for the effect of this cloud on the surface heat fluxes by coding its time of arrival at each grid point. Figure 3 shows the positions of the leading edge of the cloud at various times in the day. Two forecasts, with and without this cloud, have been carried

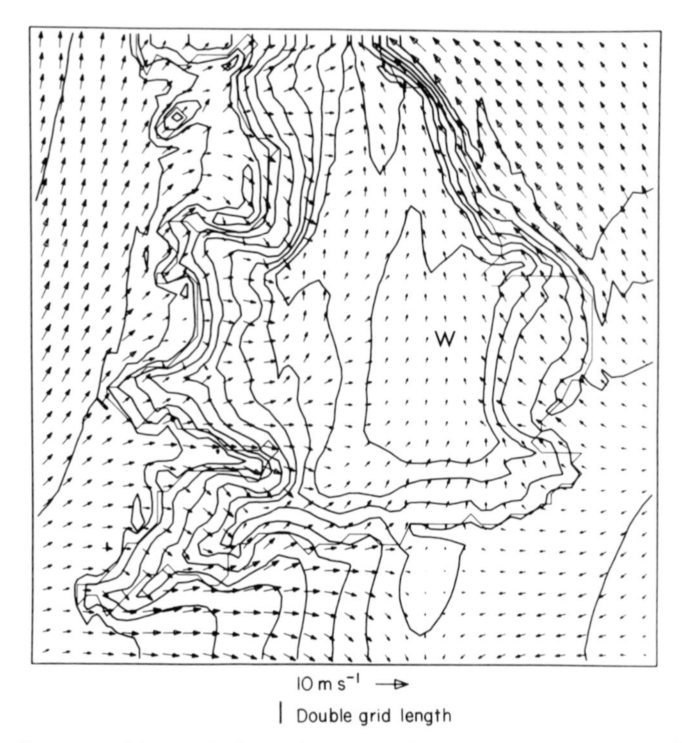

Fig. 4. Forecast 10-m winds and potential temperatures for 18 GMT on 14 August 1975 when cloud effects were excluded from the forecast model. The isopleth interval is 1 K.

out and described by Bailey *et al.* (1981). The model was modified to allow for the effect of cloud shading by halving the solar radiation that would otherwise be incident at the surface following the arrival of the cloud. The cloud had no other effect in the model, so the effect in the forecast is due solely to this cloud shading.

Figure 4 shows the forecast of low-level wind and temperature at 1800 GMT in the absence of cloud. This forecast is not inconsistent with the development of a vigorous storm over Hampstead, but as we shall see, it is strikingly different from the forecast with cloud and it seems more likely that the latter is relevant to the special events on 14 August 1975.

Figure 5 shows the forecast for 1800 GMT with the effect of the cloud edge included. Figure 6 shows the corresponding forecast ascent and also reports of thunder or cumulonimbus. The storm was probably most severe at 1800 GMT; Hampstead is shown at the southern end of the region of maximum ascent.

The correspondence between the thundery activity and forecast ascent shown in Fig. 6 is striking. At the very least, it suggests that model forecasts like this could be used as indicators of preferred areas for convective rain. We have

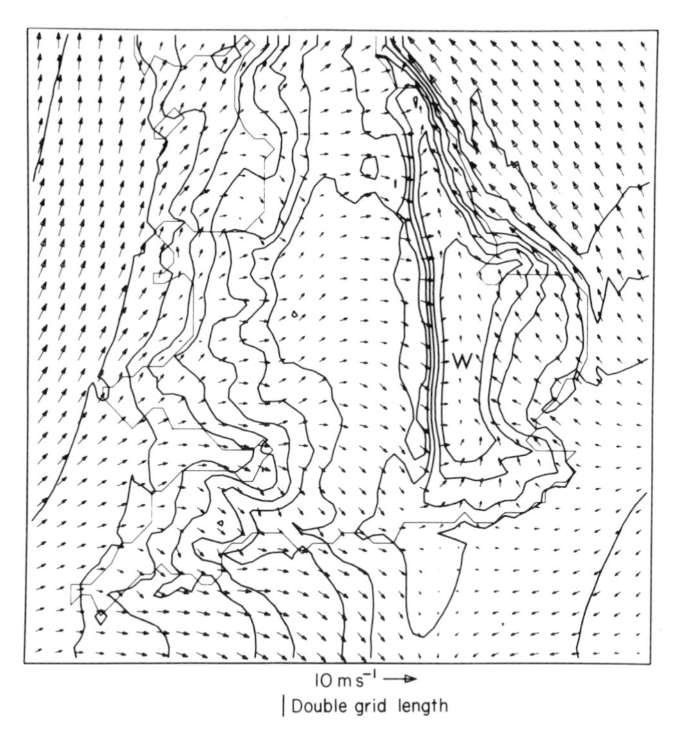

Fig. 5. Forecast 10-m winds and potential temperature for 18 GMT on 14 August 1975 when the effect of the layer cloud shown in Fig. 3 was included in the forecast model. The effect of the layer cloud can be seen by comparing these results with those in Fig. 4. The isopleth interval is 1 K.

compared the forecast with Atkinson's (1977) streamline analyses and with the routine surface reports and found reasonable, but far from precise, agreement. The problem with such a comparison is that the network of observations is not adequate to resolve the phenomena that are forecast and that it would be easy to argue that any marginal discrepancy is due to local topographic effects on the observations. In fact, we must expect some minor errors due to the lack of detail in our initial and boundary conditions.

The comparison between the forecasts with and without the inclusion of the above edge is conclusive evidence that the effects of the cloud are important, and the location of all the thundery activity close to the leading edge of the cloud (see Fig. 6) cannot be dismissed as a coincidence. Figure 7 shows a west–east section through the model, and thus the cloud edge. It can be seen that the cloud shading has induced a considerable temperature contrast throughout the boundary layer. This has in turn induced a thermally direct circulation in some

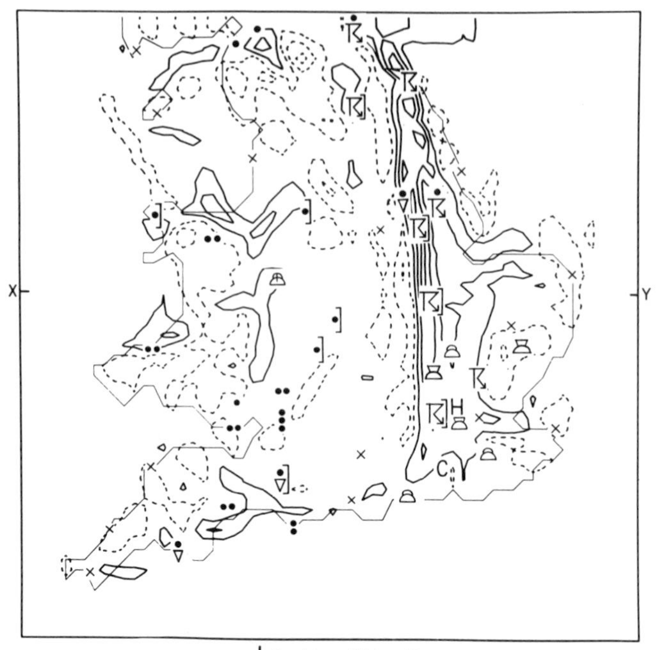

Fig. 6. Forecast vertical velocity at 190 m compared with significant weather reports, both for 18 GMT on 14 August 1975. The isopleth interval is $2\ \text{cm s}^{-1}$, which corresponds to a low level divergence of $\sim 10^{-4}\ \text{s}^{-1}$; continuous lines $+1, +3 \ldots \text{cm s}^{-1}$, dashed lines $-1, -3 \ldots \text{cm s}^{-1}$. The locations of Hampstead (H) and Crawley (C) are shown, and X and Y indicate the cross-section depicted in Fig. 7. The crosses indicate stations that failed to report thundery activity.

ways similar to a sea breeze or gravity current. It seems clear, therefore, that any reasonable local forecast model must include calculations of the shading effect of clouds. The importance of calculating the latent heat release during cloud and rain formation is familiar, but it is possibly no more important than the radiative effects of cloud. This presents modellers with a substantial problem because it is not obvious that it is easy to calculate cloud cover or cloud type with reasonable accuracy in a numerical model. In this forecast we have, in effect, treated the cloud amount as a boundary condition that must and can be obtained from some external source (in this special case, observations rather than a large-scale forecast). In operational practice, we would not have observations of clouds in advance, but it is possible that forecasts of cloud type and amount could be obtained using an objective extrapolation of nowcast cloud fields based mainly on satellite data.

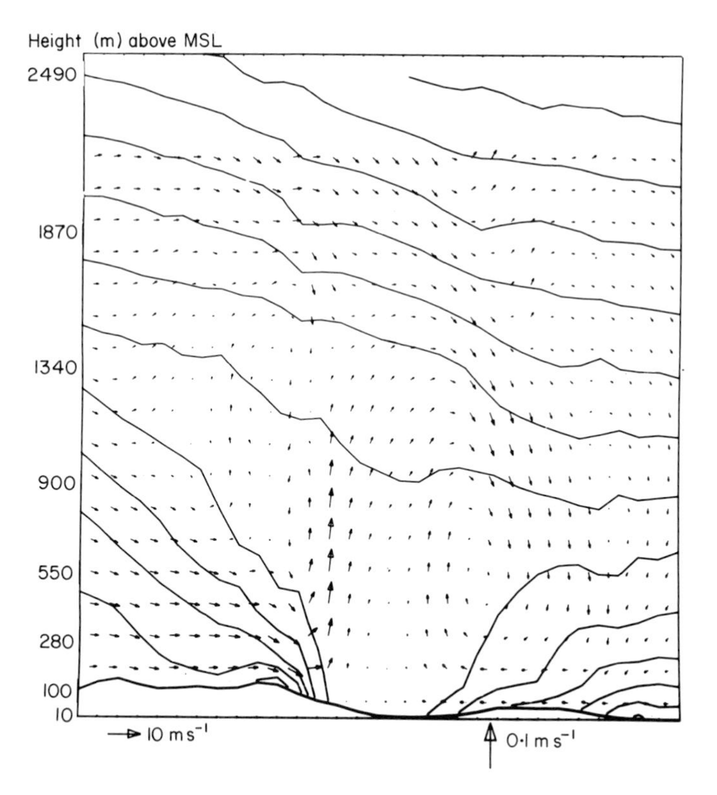

Fig. 7. Vertical west–east section (X-Y in Fig. 6) through the cloud edge showing the potential temperature and component of wind velocity in the plane of the section for 14 August 1975. The section is restricted to the bottom eight levels of the model and, approximately, to the eastern half of the model. The isopleth interval is 1 K.

5 Discussion

It is widely agreed that fine resolution mesoscale models can be expected to give good forecasts of local weather when the mesoscale meteorology is forced by local topography (Pielke, 1982). The first of the forecasts in this chapter is a straightforward example of this, although, even in this particularly simple situation, it was found that the temporal variation of the large-scale situation must be taken into account. Pielke has given several other examples. Our second forecast is also a good example of a mesoscale response to local forcing but, in this case, the forcing was due to an advancing cloud edge and not local topography. It suggests that the common expectation that mesoscale models can describe the response to local forcing covers a wider range of circumstances than is immediately obvious. A further example, which illustrates that it is difficult to know in advance where to draw the line as to what constitutes local forcing, may be the development of rain bands in association with a front. If it turns out that frontal rain bands will grow quickly in response to an imposed frontal discontinuity then, just as the Hampstead storm appears to have been forced in part by the cloud edge, the rain bands would be regarded as locally forced by the front. Thus we might expect good mesoscale model forecasts to be achievable even in the absence of good observations of the initial state of the rain bands. For these cases, the problem is in ensuring that the forcing can be prescribed accurately.

The case study in section 4 illustrates a potential advantage of combining simple extrapolation methods and numerical models. In principle one might expect to be able to forecast the existence and movement of layer cloud using the mesoscale model itself or the larger-scale model that is supplying the other boundary conditions. In practice, cloud cover is a rather special variable, not even related to cloud amount in any straightforward way, and there is no guarantee that the required accuracy can be achieved. Consequently, as noted in the previous section, it is prudent to look for an alternative way to determine the cloud cover during a forecast. The advent of geostationary satellite imagery will allow cloud cover to be monitored continuously and new interactive computer techniques (like the FRONTIERS system, Browning and Collier; 1982) will make this practicable. In principle, it would be easy to extrapolate the movement of the observed cloud through the period of a forecast and thus obtain an estimate of cloud cover that might prove more reliable for very-short-range forecasting that that generated by the models themselves.

Both the forecasts described in this chapter predict small-scale, local events with encouraging accuracy. They had relatively large lead times of about 12 hours, so the details of the initial conditions were unimportant compared with the response of the model to the large-scale situation and local orographic and topographic forcing; that is why the forecasts were good. Forecasts with shorter lead times would be less able to develop the correct response to the local forcing and the large-scale situation, and will be as good as those presented here only if the details of the initial conditions are accurate. In the absence of good initial conditions, shortening the lead time will degrade otherwise useful forecasts.

Since shortening the lead time of our forecasts could damage them, it is natural to ask whether models such as that used here can be employed to provide good very-short-range forecasts with lead times of, say, 6 hours or less. That question can be addressed properly only when we have the techniques to take advantage of the excellent but, in many respects, qualitative observations that are becoming available on the mesoscale, i.e. radar and satellite imagery, but there are reasons to be optimistic. For the locally forced situations discussed here we could produce forecasts for 6 hours from "now" as good as those shown in sections 3 and 4 by the simple expedient of starting a 12-hour forecast 6 hours before "now". Starting the forecast "now" (i.e. at the last time for which one has observations) rather than 6 hours earlier could damage the forecast in two ways, viz. by truncating the history of the synoptic-scale forcing, and by placing the model variables in a state of dynamic imbalance at the initial time. Modern data assimilation methods, which effectively allow observations to be absorbed into a continuous forecast with minimal destruction of its information content (e.g. Bengtsson, 1975) have more or less solved the first problem on the synoptic scale, and it is reasonable to expect that experience to be applicable on the mesoscale. The second problem, that of dynamic imbalance, may well prove far less damaging than might be feared because, while synoptic-scale imbalances take as long as 12 hours to adjust, the mesoscale model described here appears to adjust to an internally consistent state within about 2 hours (Carpenter and Lowther, 1982). Thus, by continuously running the model and assimilating observations as they become available, good forecasts should be possible for any time two hours or more later than the most recent observations. Since simple extrapolation methods are at their best for the shortest lead times, nowcasting and model-based forecasting complement each other and we should not be too concerned about the first two hours of a numerical forecast.

The greatest challenge facing mesoscale modellers at present is the interpretation of qualitative nowcast data in terms of the dynamical variables that control model forecasts. When this is achieved, it should be possible to derive initial conditions that are actually beneficial in all situations rather than merely harmless in situations like those discussed in this chapter.

References

Atkinson, B. W. (1977). Urban effects on precipitation: an investigation of London's influence on the severe storm in August 1975. Occasional Report No. 8. Department of Geography, Queen Mary College, London.

Bailey, M. J., Carpenter, K. M., Lowther, L. R. and Passant, C. W. (1981). A mesoscale forecast for 14 August 1975 – the Hampstead Storm. *Met. Mag.* **110**, 147–161.

Bengtsson, L. (1975). 4-Dimensional assimilation of meteorological observations. GARP Publication Series No. 15.

Browning, K. A. and Collier, C. G. (1982). This volume, pp. 47–61.

Carpenter, K. M. (1977). Surface exchanges in a mesoscale model of the atmosphere. Met. O 11, Technical Note No. 96. Copy available in the National Meteorological Library, Bracknell.

Carpenter, K. M. (1979). An experimental forecast using a non-hydrostatic mesoscale model. *Q. J. R. met. Soc.* **105**, 629–655.

Carpenter, K. M. and Lowther, L. R. (1982). An experiment on the initial conditions for a mesoscale forecast. *Q. J. R. met. Soc.* **108**, 643–660.

Keers, J. F. and Westcott, P. (1976). The Hampstead Storm – 14 August 1975. *Weather* **31**, 2–10.

Miller, M. J. (1978). The Hampstead Storm: a numerical simulation of a quasi-stationary cumulonimbus system. *Q. J. R. met. Soc.,* **104,** 413–427.
Pielke, R. A. (1974). A three-dimensional numerical model of the sea breezes over South Florida. *Mon. Weath. Rev.* **102,** 115–139.
Pielke, R. A. (1982). This volume, pp. 207–221.
Simpson, J. E., Mansfield, D. A. and Milford, J. R. (1977). Inland penetration of sea breeze fronts. *Q. J. R. met. Soc.* **103,** 47–76.
Tapp, M. C. and White, P. W. (1976). A non-hydrostatic mesoscale model. *Q. J. R. met. Soc.* **102,** 277–296.

4.3
An Example of the Use of Satellite Cloud and Surface Rainfall Data to Initialize a Numerical Weather Prediction Model

THOMAS T. WARNER, TERRY C. TARBELL AND
STEPHEN W. WOLCOTT

1 Introduction

Precipitation forecasting with numerical weather prediction models has, to date, not been very accurate, especially very-short-range forecasts in cases of significant precipitation. One of the reasons for this has been that static initializations normally employ non-divergent initial conditions. Therefore, when precipitation is occurring at the initial time, the initial precipitation rates are under-forecast since the forecast model requires several hours to develop a consistent vertical velocity field. A second reason is that mesoscale detail in the moisture field cannot be resolved by the radiosonde observations that are used as the basis for large-scale moisture analyses.

To circumvent the first of these problems, we can employ a divergent initialization procedure that can be described as follows: a vertical velocity equation that includes a diabatic term based on the observed rainfall rate diagnoses vertical velocity from observed meteorological fields. Velocity potential is derived from the vertical velocity with the continuity equation and appropriate boundary conditions. The divergent wind components are obtained from the velocity potential. The geopotentials are then calculated on sigma surfaces using a divergence equation with contributions from both the non-divergent and divergent wind. Finally, the hydrostatic temperature field is derived from the geopotential.

It has been shown (Tarbell *et al.*, 1981) that, on the mesoscale, the diabatic term in the omega equation is dominant in regions of precipitation. The observed precipitation rates, then, represent the fundamental information on which the computation of the divergent wind component is based. These precipitation rates can be obtained from either National Weather Service radar data or surface-based observations. The use of radar data would be preferred because they are spatially continuous and available in real time. One additional requirement for the technique to be successful is that the moisture initial conditions be sufficiently accurate so that saturated conditions prevail in the correct locations to allow latent heating to maintain the initial upward motion and divergence fields.

To address the second problem, the moisture analysis procedure uses three types of moisture data: radiosonde, infrared satellite imagery, and precipitation. The radiosonde observations serve as the primary source of synoptic-scale information whereas the other sources add mesoscale detail to the analysis.

The purpose of this chapter is to demonstrate the impact on very-short-range forecasts of precipitation, of using these procedures to incorporate divergent initialization and a mesoscale moisture analysis.

2 The divergent initialization procedure

2.1 SCALE ANALYSIS

The important equations used here are the omega and balance equations. The following scales are assumed: a horizontal velocity scale of $10\,\mathrm{m\,s^{-1}}$; mesoscale and synoptic-scale horizontal length scales (quarter-wavelengths) of 10^5 and 10^6 m, respectively; mesoscale and synoptic-scale advective time scales of 10^4 and 10^5 s, respectively; and mesoscale and synoptic-scale precipitation rates of 1 and $0.1\,\mathrm{cm\,h^{-1}}$, respectively. Scaling the vertical velocity equation on the synoptic-scale yields the well-known result that the important terms are the differential vorticity advection term, the Laplacian of temperature advection term, the diabatic term, and the beta term. However, on the mesoscale the diabatic term dominates the others and the beta term is negligible. Scale analysis of the balance equation simply shows that the beta term is negligible on the mesoscale but not on the synoptic scale.

2.2 BOUNDARY CONDITIONS

To use the divergent initialization procedure on the limited domain, boundary conditions are required on stream function, geopotential, omega, and velocity potential, to allow solution of Poisson equations. Anthes and Keyser (1979) analysed the effect of boundary conditions on the solution of a Poisson equation. They stated two important conclusions: (1) the influence of boundary conditions on the solution decreases exponentially with distance from the boundaries and (2) it is most important to specify accurately the large-scale (low-wavenumber) variation of the boundary conditions since amplitude errors for low-wavenumber components of the boundary conditions damp less rapidly with distance away from the boundary. We have employed various procedures for specifying the large-scale variations of these variables on the boundaries. They are discussed in detail in Tarbell *et al.* (1981).

2.3 USE OF THE OMEGA EQUATION IN THE DETERMINATION OF THE DIVERGENT WIND COMPONENT

A finite difference form of the omega equation was solved by three-dimensional, sequential over-relaxation over a 30×35 point horizontal domain at seven levels.

The finite-difference analogues of each term were constructed using centred horizontal differences. The surface friction parameterization and the procedure for determining the vertical distribution of the latent heating rate can be found in Anthes and Warner (1978) and in Tarbell *et al.* (1981), respectively.

The rainfall rates, to be used in calculation of the diabatic term of the omega equation, were based on hourly precipitation rate data obtained from a raingauge network that provided approximately one observation per $50\,km^2$. Rainfall amounts from 2 h before and 2 h after the initial time of the forecast were averaged to obtain a more representative rainfall rate. These rain rates could have been obtained from digitized radar data averaged over a short interval prior to the initial time of the forecast; however, these data were unavailable for this case. The use of radar data would have allowed the value of the precipitation rate that was assigned to each point on the domain to better reflect the appropriate average over a grid square. In an operational setting, the radar data would also be available in near real time.

3 The moisture analysis procedure

3.1 THE ANALYSIS OF RADIOSONDE MOISTURE DATA TO DEFINE SYNOPTIC-SCALE FEATURES

The radiosonde moisture analysis procedure uses a successive approximation technique that employs an elliptical weighting function to determine the influence of the observations on grid point values. Each ellipse is aligned with its major axis perpendicular to the gradient and with its centroid at an observation point. The gradients are determined from centred, finite-difference calculations made at the grid point closest to a given observation. An added option includes increasing the eccentricity of the ellipse with each pass of the data. As the length of the radius of influence is decreased after each successive pass, the ellipse becomes more elongated. This feature allows the gradient to have a greater influence as the elliptical area diminishes. This successive approximation technique can also generate its own first guess or background field, based on the actual observations, in the event that no such field is available.

3.2 USE OF PRESENT WEATHER OBSERVATIONS AND INFRARED SATELLITE IMAGERY TO PROVIDE MESOSCALE DETAIL

The moisture analysis procedure used satellite data in the form of infrared-image satellite photographs to determine total cloud volume. The cloud-top temperatures are compared with the temperature analyses at the numerical model levels. Regions of 100% relative humidity can then be determined with little ambiguity. The satellite data are applied in two different ways depending upon the present weather observations. If precipitation is observed at the location of a particular grid point, the relative humidity is assumed to be 100% from the lowest level of the model up to the observed cloud level. If precipitation is not occurring at that grid point, saturation is assumed only at the model level immediately below the observed level of the cloud top.

The specific humidity values at the grid points unaffected by the rainfall and satellite data are based only on radiosonde data outside the saturated volume of atmosphere. This approach keeps observations within the saturated region from modifying grid points outside the region and results in the creation of a strong moisture gradient at the boundary of the saturated volume.

Two cross-section analyses illustrate how the procedure works. Figure 1(a) depicts a hypothetical cloud structure along a cross-section through a cold front. Preceding the front are high, thin cirrus clouds. An anvil-shaped cumulonimbus is associated with the frontal boundary. The trailing cool dry air is interspersed with fair-weather cumuli and middle-level altocumuli. Figure 1(b) illustrates how this would appear to the numerical model after analysis. The 100% relative humidity constraint would extend to the lowest model level (level 6 in this case) for the portion of the cumulonimbus where precipitation was occurring. The altocumulus and the stratocumulus behind the front and the cirrus ahead of the front would be accurately portrayed. The small cumulus cloud preceding the cold front and portions of the cumulonimbus where precipitation was not observed do not lie within the model cloud volume. Details of the entire moisture analysis procedure can be found in Wolcott and Warner (1981).

4 The numerical model

The three-dimensional forecast model used was a modified version of the general, hydrostatic model described by Anthes and Warner (1978). The model uses the flux form of the primitive equations where the forecast variables are horizontal wind, temperature, surface pressure and specific humidity. The bulk planetary boundary layer parameterization was used.

A staggered horizontal grid is defined at six model layers between sigma levels 0.0, 0.25, 0.4, 0.55, 0.7, 0.85, and 1.0. The pressure at the top sigma level was 250 mb and a time step of 180 s was employed for the grid increment of 120 km.

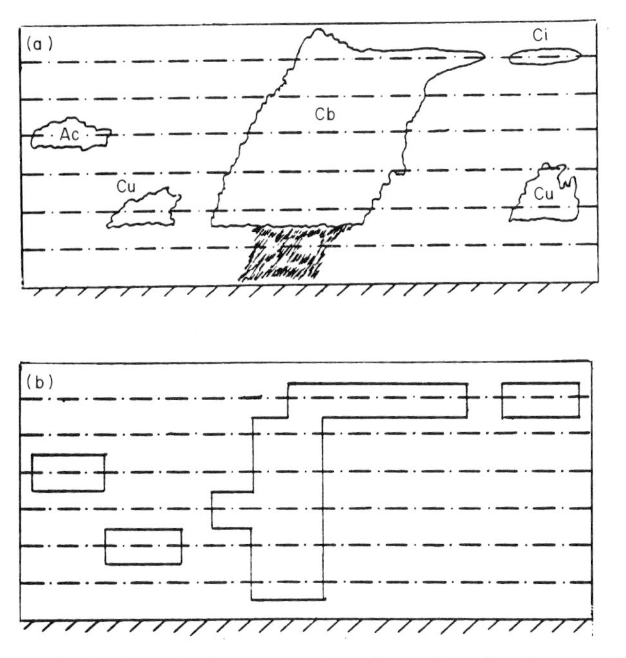

Fig. 1. A hypothetical, vertical cross-section through a cloud system (a) and the results of applying the relative humidity constraints based on infrared-image satellite photographs and precipitation observations (b). The dots at each of the six model levels represent grid points.

The 30 x 35 grid was centered at 40°N, 95°W. The model employs a time smoother as well as enhanced horizontal diffusion near the boundaries. Both convective and non-convective precipitation are forecast.

5 The case study

5.1 SYNOPTIC SETTING

A high pressure area persisted over the southeastern United States between 1200 GMT 19 November and 0000 GMT 20 November 1975. The ridge provided a continuous supply of low-level moisture from the Gulf of Mexico to the Great Plains and adjacent states. This moisture supply was a key factor in the precipitation which occurred during this 12-h period.

The observed sea-level pressure at 1200 GMT 19 November 1975 is presented in Fig. 2. The ridge dominates eastern United States while a trough extends from a 1008-mb low in Texas through Minnesota. Relatively rapid changes occurred in the following 12-h period. A low developed over the Texas panhandle and

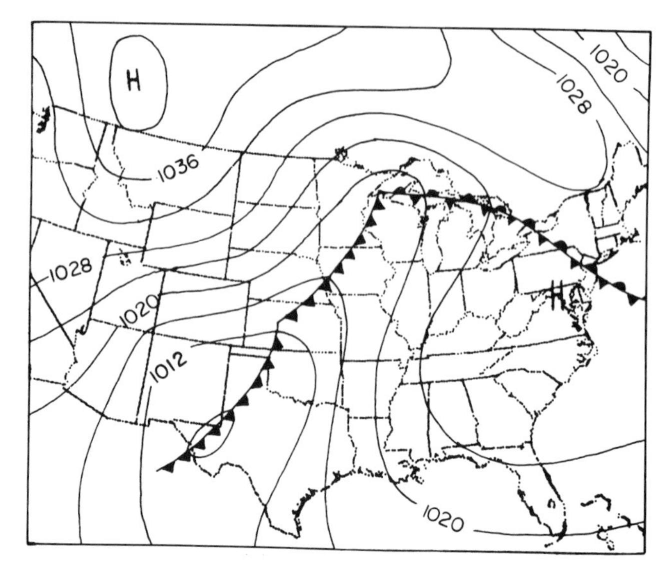

Fig. 2. Observed sea-level pressure field at 1200 GMT 19 November 1975. Surface fronts are indicated and the contour interval is 4 mb.

moved to eastern Kansas as well as deepening rapidly to 1004 mb. The trough extended from Texas through the Kansas low and then through Lake Huron and into New England. The surface pressure gradient from Wyoming to Kansas strengthened considerably.

5.2 INITIAL CONDITIONS FOR THREE FORECASTS

We will discuss three 12-h forecasts, all beginning at 1200 GMT 19 November 1975, which will illustrate the benefits of the divergent initialization and the high resolution moisture analysis. The first two forecasts utilized all the sources of moisture data discussed, but differed in terms of their initial divergent wind field. Forecast 1 was initialized with non-divergent winds and balanced temperatures as described by Warner *et al.* (1978), while forecast 2 employed the divergent initialization procedure. Forecast 3 employed the divergent initialization but used an initial moisture analysis based only on radiosonde data. Table 1 summarizes the three simulations.

Figure 3 illustrates the observed rain rate at the initial time of the forecast. Use of these data in the omega equation and subsequent calculation of the divergence fields provides the divergent wind components. The computed divergent wind field for the 925-mb level is given in Fig. 4. There is a narrow zone of convergence from central Wisconsin to southwestern Kansas and on toward the

TABLE 1
Summary of forecast initial conditions

Forecast	Data used in moisture analysis	Initial wind field
1	Radiosonde, satellite, precipitation	Non-divergent
2	Radiosonde, satellite, precipitation	Divergent
3	Radiosonde	Divergent

south. This low-level convergence is supported by divergence aloft (not shown). The initial relative humidity field for the sigma level located near 830 mb is shown in Fig. 5. The satellite picture used in the moisture analysis procedure is illustrated in Fig. 6.

5.3 COMPARISON OF FORECAST RESULTS

The accuracy of the three forecasts summarized in Table 1 will be compared only in terms of the precipitation predictions. The other predicted fields did not show a strong sensitivity to the use of the high resolution moisture analysis or the initial divergent wind field; further discussion of this subject is found in Tarbell *et al.* (1981) and Wolcott and Warner (1981).

Figure 7 illustrates the total precipitation volume for each hour through the forecast period for each simulation as well as the total volume of observed precipitation. The total amount of precipitation forecast by the model is a useful statistic because the forecasts over the whole domain can be compared. It is clear that forecast 2, which used the high-resolution moisture analysis and the divergent initialization, produced a substantial improvement over the other forecasts during the first 4 hours.

Figure 8 shows the spatial distribution of the precipitation during the first 3 h based on observations, as well as for forecasts 1, 2, and 3. The first point to note is that all the forecasts gave smaller precipitation amounts than those observed. However, comparing (b) and (c) in Fig. 8 indicates that significantly more precipitation occurred with divergent initial conditions than with non-divergent initial conditions when the high-resolution moisture analysis was employed. Figure 8(d) shows that the divergent initialization produced an organized and realistic pattern of precipitation even when a conventional moisture analysis was used. However, the amounts were not as realistic as in Fig. 8(c). These results apply to the first 3 hours of the forecast; smaller but still significant improvements in the 12-h rainfall totals are also apparent when divergent initial conditions are used with the high-resolution moisture analysis.

Because the rainfall pattern in this case remained relatively stationary after the start of the forecast, persistence provides a better rainfall prediction than

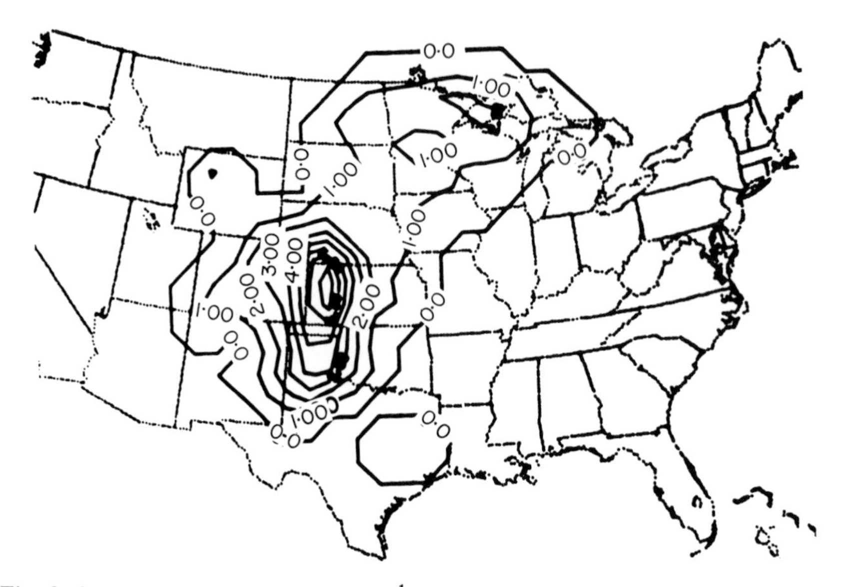

Fig. 3. Observed rainfall rate (cm day^{-1}) valid at 1200 GMT 19 November 1975. This rate was derived from the observations for the period 1000 to 1400 GMT 19 November 1975. The contour inverval is 1 cm day^{-1}.

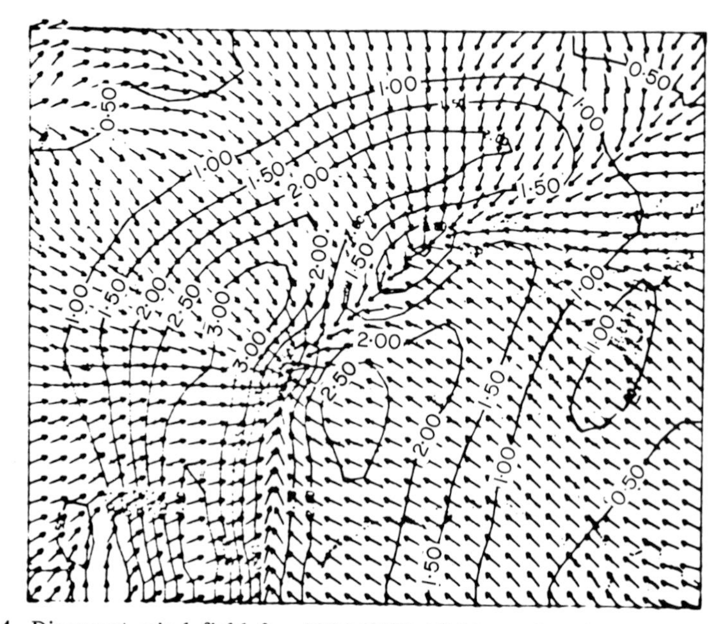

Fig. 4. Divergent wind field for 1200 GMT 19 November 1975 at 925 mb as implied by the observed rainfall field in Fig. 3. The contour interval is 0.5 m s^{-1}.

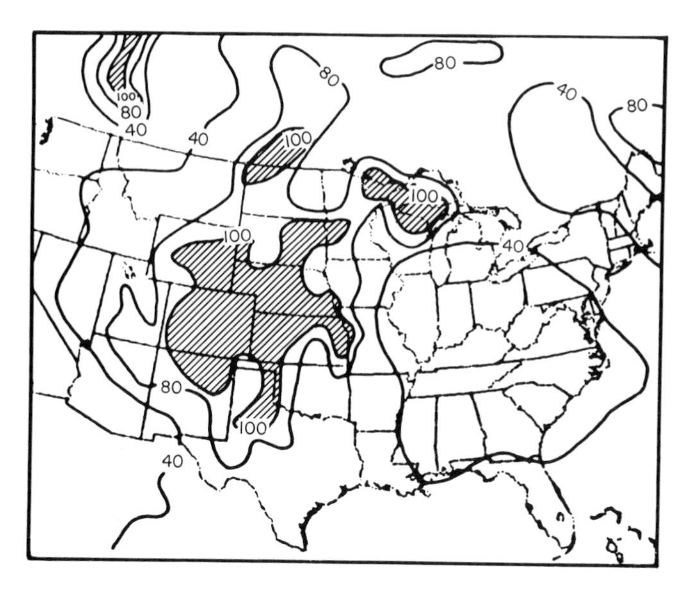

Fig. 5. Relative humidity analysis valid at 1200 GMT 19 November 1975 for sigma level $5\frac{1}{2}$ ($p \approx 830$ mb). The contours represent 40, 80 and 100% relative humidity.

Fig. 6. Infrared-image satellite picture of the domain taken at 1145 GMT 19 November 1975.

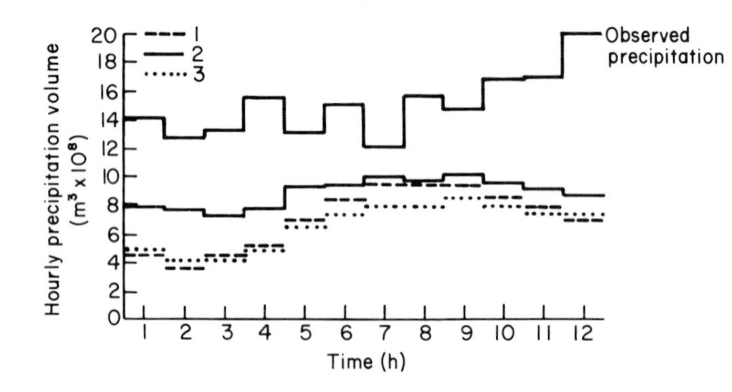

Fig. 7. Total hourly precipitation volume for the entire domain for forecasts 1 to 3.

any of the model forecasts. However, in cases where the rainfall pattern is undergoing rapid transitions in intensity or direction of propagation, the modelling approach would have a definite advantage over persistence for forecast times of longer than say a few hours.

6 Conclusions

For this case, divergent initializaton significantly improved the very-short-range forecast of precipitation with a numerical weather prediction model started from a static initialization. However, a moisture analysis procedure that specified saturated conditions based on satellite cloud imagery as well as surface-based rainfall observations was also required in order to produce the greatest improvement in the precipitation forecast. Greater differences in forecast performance are likely for cases with higher precipitation rates at the initial time; this would be because of the even poorer performance of forecasts based on non-divergent initial conditions in such circumstances.

It is interesting to speculate on how much improvement could be expected in rainfall forecasts if these initialization techniques were applied in conjunction with a model of considerably finer resolution, i.e. a truly mesoscale model. The precipitation patterns produced by a finer-scale model should potentially be more localized and intense and thus the use of initial conditions that contain fine-scale moisture and wind information would seem to be essential. In addition, the magnitude of the divergent wind on the smaller-scale would be larger and thus its omission from the initial conditions could be more detrimental to the forecast.

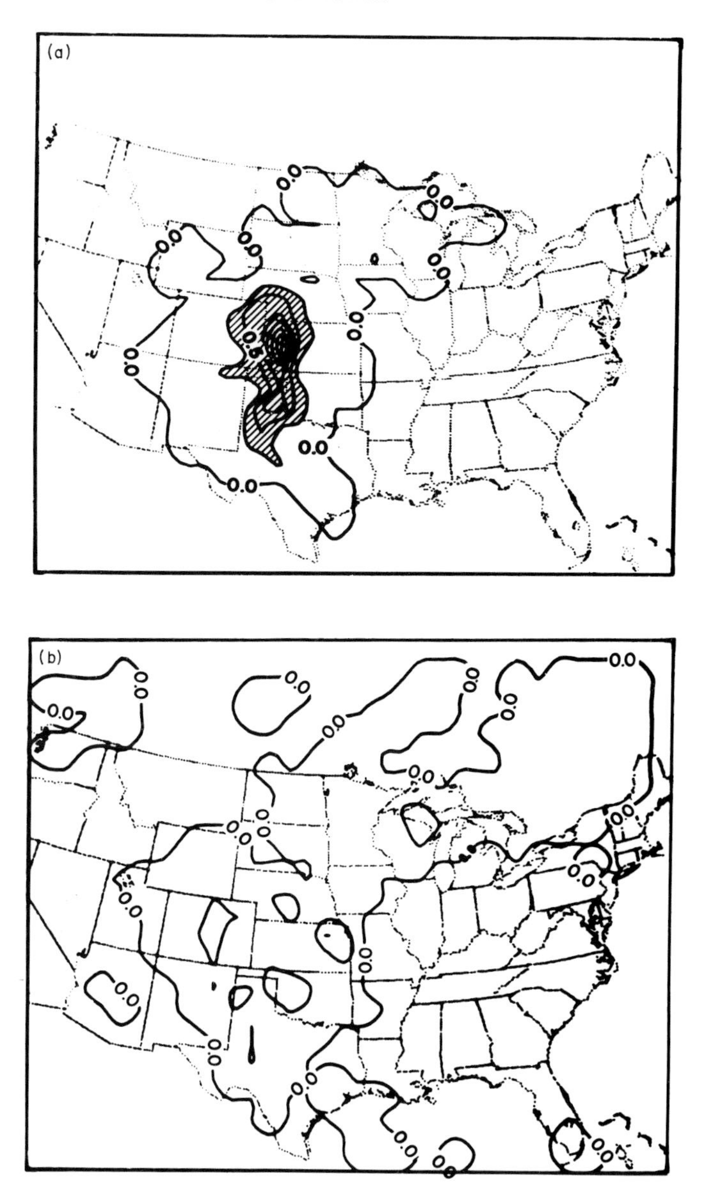

Fig. 8. Observed precipitation for the first three hours (1200–1500 GMT 19 November 1975) of the forecast period, and the forecast precipitation for the same period for forecasts 1 to 3.

(a) Observed precipitation. The contour interval is 0.25 cm.

(b) Forecast 1 based on a high-resolution moisture analysis with a non-divergent initialization. The contour interval is 0.125 cm.

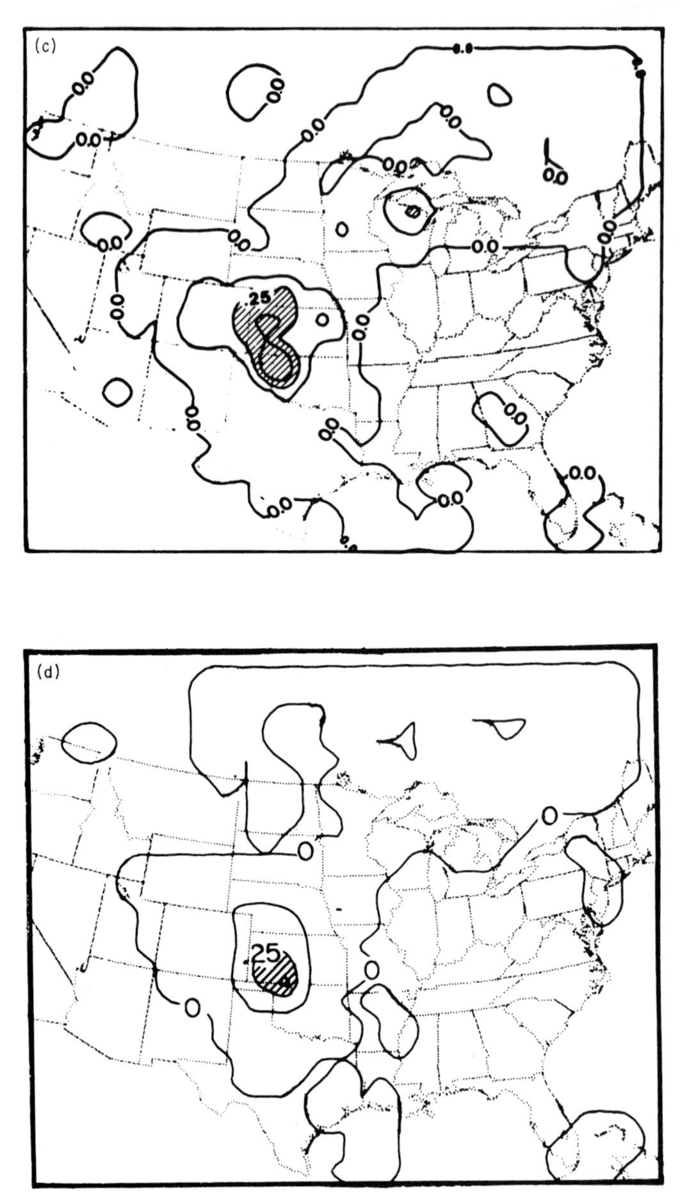

(c) Forecast 2 based on a high resolution moisture analysis and the divergent initialization. The contour interval is 0.125 cm.

(d) Forecast 3 based on a conventional moisture analysis coupled with the divergent initialization. The contour interval is 0.125 cm.

Acknowledgments

This research was made possible under Environmental Protection Agency Grant R805659 and National Aeronautics and Space Administration Grant NSG5205. The computer facilities of The Pennsylvania State University and the National Center for Atmospheric Research (NCAR), Boulder, Colorado, were used for this research. NCAR is sponsored by the National Science Foundation.

References

Anthes, R. A. and Keyser, D. (1979). Tests of a fine mesh model over Europe and the United States. *Mon. Weath. Rev.* **107**, 963–984.

Anthes, R. A. and Warner, T. T. (1978). The development of mesoscale models suitable for air pollution and other mesometeorological studies. *Mon. Weath. Rev.* **106**, 1045–1078.

Tarbell, T. C., Warner, T. T. and Anthes, R. A. (1981). An example of the initialization of the divergent wind component in a mesoscale numerical weather prediction model. *Mon. Weath. Rev.* **109**, 77–95.

Warner, T. T., Anthes, R. A. and McNab, A. L. (1978). Numerical simulations with a three-dimensional mesoscale model. *Mon. Weath. Rev.* **106**, 1079–1099.

Wolcott, S. W. and Warner, T. T. (1981). A moisture analysis procedure utilizing surface and satellite data. *Mon. Weath. Rev.* **109**. In press.

Subject Index